PROFESSIONAL ETHICS
AND CIVIC MORALS

In this book Durkheim outlined the core of his theory of morality and social rights which was to dominate his work until his untimely death in 1917. Durkheim saw sociology as a science of morals which are objective social facts; these moral regulations form the basis of individual rights and obligations. The book is crucial for understanding Durkheim's sociology, because it contains his much neglected theory of the state as a moral institution. It is also essential for understanding his critique of anomie and egoistic individualism.

The growing interest in cultural relations and moral regulation associated with recent contributions in historical sociology, makes this new edition of Durkheim's classic work especially timely. It shows that Durkheim had worked out a position on the modern state which is a genuine rival to the Marxian and Weberian traditions. Durkheim's stress on the moral regulation of everyday life chimes with current concerns with individual freedom and the contours of permissible behaviour. It is an essential resource in understanding the state and society and it can also be read as a crucial work in modern social theory.

Bryan S. Turner has done a superb job in showing the social and political influences on Durkheim during the writing of the book, notably the notorious Dreyfus affair. He also shows how the book should be interpreted in relation to other key works in Durkheim's *oeuvre* and its relevance for modern sociology.

ROUTLEDGE SOCIOLOGY CLASSICS

Editor: Bryan S. Turner

FROM MAX WEBER
*Translated, Edited and with an
Introduction by*
H.H. Gerth and C. Wright Mills

IDEOLOGY AND UTOPIA
Karl Mannheim

THE SOCIAL SYSTEM
Talcott Parsons

PROFESSIONAL ETHICS
AND CIVIC MORALS

by

EMILE DURKHEIM

Translated by
Cornelia Brookfield

With a new preface by
Bryan S. Turner

LONDON AND NEW YORK

First published in the English language in 1957
by Routledge
11 New Fetter Lane, London EC4P 4EE

Simultaneously published in the USA and Canada
by Routledge
a division of Routledge, Chapman and Hall Inc.
29 West 35th Street, New York, NY 10001

Printed and bound in Great Britain
by Mackays of Chatham PLC, Kent

British Library Cataloguing in Publication Data
Durkheim, Émile 1858–1917
Professional ethics and civic morals – 2nd ed. – (Routledge
classics in sociology).
1. Ethics
I. Title   II. Leçons de sociologie physique des moeurs et du droit
170

Library of Congress Cataloging in Publication Data
Also available

ISBN 0–415–06225–X

TRANSLATOR'S NOTE

As earlier translators of Durkheim have found, rendering the text in English requires interpretative treatment. Durkheim is often inclined to anthropomorphism, which is carried by the genders of French substantives. In English, fidelity to such liveliness would fail in its purpose and I have given such passages a more sober turn. Further, these Sorbonne lectures exist in the original only as Durkheim's personal Notes or working transcription of those Notes. The English version attempts to preserve a tone of the spoken word. And the repetitions of one or two passages have been left as found in the University of Istanbul publication. Where it has been possible to trace Durkheim's references (clearly for his own use in lecturing to his students) to volumes on his own library shelves, I have completed his mere indications to serve the English-speaking student. Finally, philosophic terms have been rendered as consistently as possible, after consultation with specialists in the various subjects discussed by the Author.

In regard to Professor Georges Davy's Introduction, there has been no escape for the translator in those abstruse excursions on to the high plateau of philosophic speculation. Here, fidelity has stood in the way of "plain" English.

C.B

CONTENTS

PREFACE

BY PROF. H. N. KUBALI

THIS work, brought out by the Faculty of Law in the University of Istanbul, is a collection of some hitherto unpublished lectures of Emile Durkheim.

In 1934, in Paris, I had set about the writing of a thesis for a doctorate in law on "The Concept of the State held by the Pioneers in the French School of Sociology". It then seemed to me that before all else I must make a study of the precise ideas and thought of Durkheim, as founder of this school, on the problem of the State.

As a sociologist, Durkheim had made no special study of this problem and was satisfied in his published works merely to raise certain questions relative to it. On that account I came to the conclusion that relevant and detailed exposition might perhaps be found in his unpublished work, if any such existed. To this end I approached the well-known ethnographer Marcel Mauss, the nephew of Durkheim. He received me most cordially and spoke of his great feeling for Turkey, which he had visited in 1908. He then went on to show me a number of manuscripts with the title "The Nature of Morals and of Rights". These, he said, were a course of lectures given by Durkheim between the years 1890 and 1900 at Bordeaux and repeated at the Sorbonne, first in 1904, and then in 1912, and revived in lectures some years before his death. Mauss had no hesitation in entrusting them to me, a fact which I recall with pleasure, and he handed over to me at my request a typescript copy of part of the manuscripts likely to be of especial interest to me. I take this opportunity of paying tribute to the memory of the late scholar: I owe him a debt for his invaluable help.

Mauss had told me at the time of our talk that he intended to publish these manuscripts in *Les Annales Sociologiques*, he being a member of the editorial committee. But he only published the first part of them, made up of the three lectures on

ix

Professional Ethics, and this was in 1937, in the *Revue de Méta-physique et de Morale*. He did this, as he writes in his introductory notes, to comply with the instructions given a few months before his death in 1917 by Durkheim, who intended some of his manuscripts for Xavier Léon, the founder of the *Revue de Métaphysique et de Morale*, in preference to others, as a mark of his friendship. In doing this, Mauss announced that he would publish later with these three lectures those on Civic Morals which followed them.

In 1947, I published a Turkish translation of six lectures on Civic Morals, which I had at my disposal, in the *Revue de la Faculté de Droit d'Istanbul*. I had seen no trace of the publication planned by Mauss but I wanted first to make sure whether it had been done. I enquired, but there was no reply from him. I then, with the help of M. Bergeaud of the French Embassy, appealed to Durkheim's daughter, Madame Jacques Halphen. Mme. Halphen kindly sent me word that Marcel Mauss was exhausted by all that he had suffered during the Occupation and was not in a fit state to give any details at all. She later let me know that she had been able to identify the manuscripts in question from the copy I had sent her and that they were now in the Musée de l'Homme with all the books and papers of the Marcel Mauss Collection. Besides the three lectures on Professional Ethics already published, these manuscripts included, as she told me, fifteen lectures on Civic Morals which had not so far appeared in France.

Some months later I considered getting the whole of these lectures published through the Faculty of Law of Istanbul. I consulted Mme. Halphen and she readily agreed to the plan, which the Faculty was pleased to approve.

Such are the circumstances in which the manuscripts came to light. According to Mauss, in the *Revue de Métaphysique et de Morale*, they form the sole text of a final draft made between November 1898 and June 1900 and are now published in this volume. These facts, too explain how the plan I had at heart came to be carried out successfully.

I must therefore, before going further, express to Mme. Halphen the deep gratitude of the Faculty of Law of Istanbul as well as my own, for so kindly giving permission to us to bring out this unpublished work of her famous father. I must also give warm thanks to my very distinguished colleague, Monsieur Georges Davy for agreeing to undertake the difficult task of giving the finishing touch to the manuscripts and for writing an introduction. As a disciple and friend of Durkheim, no one had greater authority than Monsieur Davy, as an eminent sociologist, to give us this valuable help. I also want to give very especial thanks to Monsieur Charles Crozat, Professor in our Faculty, as well as to Monsieur Rabi Koral, Reader in the same Faculty, for reading the proofs and for giving such great care to seeing the book through the press.

The publication in Turkey of this posthumous work of Durkheim is not in any way a matter of chance but rather, we might say, the result of a kind of cultural determinism. For in Turkey, Durkheim's is the only sociology, apart from that of Le Play, Gabriel Tarde, Espinas and others, to have become a standard work, especially since the books and teaching of Ziya Gökalp, the well-known Turkish sociologist. There are many like myself in Turkey who bear the stamp of Durkheim's school of thought. It is therefore not surprising that Turkey, if I may say so, feels she shares a right to the heritage of this school. On this score, the country welcomes the publication with just pride. She certainly truly appreciates a fact without precedent—that the unpublished work of a European scholar of world-wide reputation is brought out here, through the good offices of one of her learned institutions.

For its own part, the Faculty of Law of the University of Istanbul has every right to be proud of having thus helped to strengthen the traditional ties of culture and friendship between Turkey and France. It is equally proud of having helped to enrich the heritage of learning common to both nations by securing the publication of a work of this distinction, and of having thus at last paid the tribute it owes to the memory of Emile Durkheim.

I feel a profound satisfaction in having had a modest part in setting this plan on foot, and I am glad to have done this service for my own country and to have been the means of spreading the light of French learning, to which I myself owe so much.

Hüseyin Nail Kubali,
Dean of the Faculty of Law of Istanbul.

xi

PREFACE TO
THE SECOND EDITION

Bryan S. Turner

INTERPRETING EMILE DURKHEIM

EMILE DURKHEIM (1858–1917) remains a major figure in social science as a whole and he is unambiguously a 'founding father' of sociology. Whereas other social theorists from the classical period of sociology (1890–1920) were often somewhat ambiguous about their status as 'sociologists', Durkheim appears to have had a clear vision of the importance of building sociology as a science of social facts. His sociology continues to play a profound role in shaping contemporary thought about the nature of modern life, and anybody who wants to understand modern French social thought must take Durkheim seriously. His work remains a rich and challenging resource for comprehending the complexity of the modern world, a complexity which Durkheim described, by adopting the moral philosophy of Jean Guyau (Orru 1987) as 'anomic'. Unlike other dominant figures who have shaped modern social theory (such as Georg Simmel, Max Horkheimer, or Talcott Parsons) and who often wrote in a dense and often obscure prose, Durkheim's writing is direct, concise, and comprehensible. His books often start with a difficult analytical problem such as the meaning of 'religion' in the opening sections of *The Elementary Forms of the Religious Life* (Durkheim 1961), but his arguments are invariably logical and clear. From the point of view of a student of sociology, Durkheim is in this sense an accessible author. Yet the clarity may be deceptive, because the underlying problems of Durkheimian sociology – can one have a *science* of morals? – are clearly immense.

The style and contents of *Professional Ethics and Civic Morals* are, in this sense, typical. Durkheim's purpose was to explore the moral problems of an advanced, differentiated, and complex society, in which the economy had become somewhat detached from other social institutions. Much of the text is concerned to

xiii

establish a clear analytical understanding of major concepts (sanction, property, morals, and contract), but this search for definitional clarity in order to remove the misconceptions of existing theories prepares the way for Durkheim's major concern, which was: how can we find a system of moral restraint which is relevant to modern conditions? The answer was, at least in part, in terms of the evolution of systems of professional codes and civic values, which would contribute to a regulation of the economy rather as the guilds had regulated medieval economic activity (Black 1984). The state, which Durkheim saw as part of the moral apparatus of society, had an important part to play in regulating social life, but also, as we will see, in protecting the rights of the individual. This answer also provided a sketch of his sociology as a whole, which was, for Durkheim, essentially a science of morals.

Although the style and the content of the argument appear at this level to be relatively simple, Durkheim's sociology has been surrounded by a forest of contradictory and often misleading interpretation. Before turning to the thesis which is embedded in *Professional Ethics and Civic Morals* (hereafter *Civic Morals*), we need to understand some of the principal exegetical frameworks within which Durkheim's work has been received, especially in the English-speaking world. This overview of the tradition of interpretation is important, because I wish to argue that *Civic Morals* is a challenge to these paradigms of interpretation and reception. In particular, it is important to question two conventional views of Durkheim's sociology. The first is that his work is, in some sense, conservative, because it was primarily concerned to understand social order rather than social change, and the second is the claim that there is a major break between his early and his later sociology. I shall address these issues in this order.

FRENCH SOCIETY (1789–1918)

Between 1789 and 1914, France was subject to profound revolutionary changes which not only transformed French society but, in a real sense, created 'modern society' as a global phenomenon. The French Revolution and the Napoleonic period experimented with and then exported the elementary principals of modern democracy, namely liberty, equality, and fraternity (or secular solidarity). The destruction of the *ancien regime* resulted, however, in The Terror, and produced throughout Europe a conservative reaction against the excesses of the liquidation of the aristocracy and the monarchy. Perhaps the most famous response in the English-speaking world was Edmund Burke's *Reflections on the Revolution in France* which became, possibly in contradistinction to Burke's own ideas, a manifesto against revolution. Jeremy Bentham in his *Anarchical Fallacies* called the idea of natural rights in the 'Declaration of the Rights of Man and the Citizen', with his characteristic vigour, 'nonsense upon stilts' (Waldron 1987: 53).

The period between the Second Restoration (1815), the death of Napoleon (1821), and the Revolution of 1848 was marked by various unsuccessful attempts to create a stable government under a constitutional monarchy (Cobban 1961). Marx in *The 18th Brumaire of Louis Bonaparte* wrote rather contemptuously of these political struggles as a 'farce' (Feuer 1969: 360). However, the 1848 Revolutions throughout Europe raised once more the hope of a liberal, bourgeois alternative to the reactionary regimes which ruled over European affairs after the fall of Napoleon. The failure of the 1848 Revolutions, especially in France and Germany, was the context in which conservative social forces were able to maintain their traditional political role, despite the industrialization of Europe which placed considerable economic power in the hands of the urban bourgeoisie, which embraced various combinations of reformism, nationalism, and liberalism.

French society was further brutally transformed by military defeat in the Franco–Prussian War of 1870, in which Alsace-Lorraine, the birth-place of Durkheim and the focal point of a strong Jewish community, was annexed by Prussia. Military failure contributed to growing social tensions between social classes, and between Catholic conservatism, nationalism, and anti-Semitism, on the one hand, and liberal, secular, bourgeois groups, on the other. In France, these conflicts resulted eventually in the bloody confrontation of the Paris Commune of 1871. Karl Marx and Friedrich Engels, observing these events from London, expected an immediate, devastating, and final revolutionary struggle by the working class against the oppression of the capitalist system. Their revolutionary aspirations were soon dashed by the bloody suppression of the Commune.

The constitutional laws of 1875, which consecrated the Third Republic, emerged out of this traumatic period, but it did not provide a solution to the political divisions in France between a traditional Catholic political bloc and radical secular socialism. In this sense, the politics of the nineteenth century in France was an attempt to come to terms with the legacy of the French Revolution, and to settle the struggles between monarchy, republicanism, and Bonapartism within an effective constitutional framework. Military defeat in 1870 produced a deep nationalistic response in which the French population, including the intelligentsia, desperately sought a regeneration of the nation (Lukes 1875: 41). In fin-de-siècle France, there was a significant wave of anti-Semitism, which had its parallel in most of the major cultural centres of Europe, but especially in Vienna. Jews were thought to be unpatriotic, but they were also assumed to be secular rationalists and therefore anti-clerical. They were, according to anti-Semitic mythology, simultaneously a threat to the state and the church. These tensions were the backcloth to the famous 'Dreyfus Affair' (1894) which divided the French nation for over a decade (Miquel 1968). Captain Alfred Dreyfus, an Alsatian Jew from a wealthy family, was accused of selling official military secrets to the Germans; he was eventually charged and convicted of treason. Knowing himself to be innocent, Captain Dreyfus failed to obey the code of military gentlemen by refusing to commit suicide or to confess. He was sentenced to life imprisonment on Devil's Island (Fenton 1984: 14), but the case remained stubbornly open and contested. After a retrial and a presidential pardon, the Dreyfus case was finally closed by the Appeal Court in 1906.

The Affair further divided French society into Catholic, conservative nationalists and secular liberals and radicals. Much of the emotional fervour of the anti-Dreyfusards was directed against 'intellectuals' who were held to be a corrupting force in French society. It was in the context of that attack that Durkheim wrote his 'Individualism and the intellectuals' (Durkheim 1969) for *La Revue Bleue* in 1898. Durkheim, who came from an established rabbinical family, was, as a university professor, inevitably caught up in the Affair, especially after a local newspaper in Bordeaux had suggested that Durkheim had encouraged his students to become politically active. Emile Zola's letter 'J'accuse' which was addressed to the President of the Republic in January 1898, accused the officers and judges who directed the case against Dreyfus of incompetence and prejudice. Zola's letter intensified the polarization between intellectuals and conservatives. Durkheim's attitude towards the Affair is revealing. He wanted to avoid clouding the issue with conflicts over politics and personalities. For Durkheim, the Affair was a moral rather than political turning-point in the history of the nation. The case, which was in reality a legal farce, was in Durkheim's opinion an opportunity for national renewal.

France was further devastated in the catastrophe of the trenches of Normandy in 1914–1918. This national tragedy was also a personal disaster for Durkheim, many of whose intellectual disciples were slaughtered in the war. Over 30 per cent of the students from the Ecole Normale Supérieure who went to the firing line were destroyed. Durkheim wrote two pamphlets in connection with the war: *Qui a voulu la guerre?* (Durkheim 1915a) and *L'Allemagne au-dessus de tout* (Durkheim 1915b). Unfortunately, even during the war Durkheim, a Jew with a German name, came under criticism. His son André was killed in the Serbian retreat of 1915–16 (Giddens 1978: 20). André Durkheim was a member of the intellectual community which had gathered around the journal *Année sociologique* which Durkheim had founded in 1896 (Nandan 1980). His death was simultaneously a personal and intellectual tragedy. As a result of exhaustion and grief, Durkheim eventually succumbed to a stroke and, after a brief recovery, died at the age of 59.

CONSERVATISM AND SOCIOLOGY

The origins of not only French, but of classical European, sociology have to be understood in the context of these profound social and political crises. Robert Nisbet (1967) in *The Sociological Tradition* has argued that sociology was an aspect of diverse intellectual movements which were responses to the industrial and the French Revolutions. This sociological response was filtered through three doctrines: socialism, conservatism, and liberalism. However, the most significant force shaping early sociology was in fact conservatism. The key ideas or 'unit ideas' of sociology, such as the problem of authority, the sacred, community, the problem of the individual, status in relation to social change, and organic wholeness are primarily aspects of this conservative intellectual legacy. Thus, sociology was an intellectual response to the sense of a lost community, the disappearance of the sacred as a source of values, the isolation of the individual in the city, and the resulting crisis of meaning. In this sense, sociology was a nostalgic reflection on the loss of authenticity, personal spontaneity, social wholeness, and community (Stauth and Turner 1988). Ferdinand Tönnies's famous distinction between *gemeinschaft* (community) and *gesellschaft* (association) (Tönnies 1957) was a crucial contribution to the subsequent idea that modern societies are fragile and superficial, because they are not grounded in lasting values.

How did Durkheim stand, according to Nisbet (1967), within this tradition? Although Durkheim's search for a rational and positivistic theory of morals was a legacy of the Enlightenment project, Durkheim adopted and developed five themes which were derived essentially from a conservative tradition. These conservative themes were: the primacy of society over the individual; the necessity for moral restraint over human passions; the importance of authority in the organization of communities; the dependence of society on religious values; and the organic character of social relations. It is important to consider each theme in order to grasp fully the argument that Durkheimian sociology was part of a conservative reaction to social change. In order to clarify this presentation, it is important to note that, while there is much to commend Nisbet's interpretation, I shall eventually depart decisively from his exegesis to offer an alternative view of Durkheim.

Durkheim criticized the liberal and utilitarian traditions by arguing that 'society' is ontologically prior to the 'individual'. For example, in *The Rules of Sociological Method* (Durkheim 1964), Durkheim defined sociology as the scientific study of social facts which are to be treated as things, that is social phenomena which exist independently of the subjective appraisal of individuals. Social facts are *sui generis*. Although this approach to sociology has often been condemned as positivistic and inadequate, it is possible to provide a defence of Durkheim's account, if we realize that he was not trying to define the research methods which sociologists are to employ in routine sociological inquiry (Gane 1988). Durkheim was also trying to offer a method of 'reading' social facts which would avoid ideological and personal bias. By 'a social fact', Durkheim meant social phenomena which are external to an individual and which exercise a social or moral constraint over behaviour. Social facts include such phenomena

xviii

as legal institutions, religious belief systems, and financial systems; they also include 'social currents' (Durkheim 1964: 4) or what we would now term 'social movements'. The data of *Civic Morals* (legal sanctions, moral codes, customs, and so forth) are social facts in Durkheim's terms. The 'rules' of sociology attempt to outline how true knowledge of these social facts might be produced. Now Nisbet takes this treatment of the relationship between the individual and society in sociological methods as an example of conservatism, because the 'ideas, language, morality, and relationships' of an individual 'are but reflections of the anterior reality of society' (Nisbet 1965: 25).

Second, human nature is such that moral constraint is essential for the well-being of humans and for the stability and safety of society. As Nisbet points out, the Enlightenment tradition saw Man as a creature of almost infinite capacity, whose nature had been stunted by religious control, political tyranny, or social corruption. As Rousseau had argued in *The Social Contract*, Man is born free, but everywhere he is in chains. By contrast the conservative tradition, especially under the influence of the Christian doctrine of the sinfulness of Man, regards human beings as creatures who need discipline in order to regulate their desires. We can take one famous example of this form of reasoning in Durkheim in his study of suicide (Durkheim 1951), where the idea of anomie plays a pivotal role.

Durkheim adopted a view of Man which is best described as 'homo duplex'. Rather like the famous story of Jekyll and Hyde, human beings have two opposed natures. One is violent and passionate; the other is rational and sociable. The requirements of social stability demand the subordination of the animality of human beings by reason, if society is to avoid anarchy. Theories of society which are based on the assumption of 'homo duplex' typically argue that, whatever the individual cost, human sexuality must be regulated in the interests of social order. Sigmund Freud's treatment of this issue can be found in *Civilization and its Discontents* (Freud 1930). For Durkheim, the problem of modern society is that, with the decline of the principle of mechanical solidarity which is based on a shared system of beliefs and morals (that is on the *conscience collective*), human beings are exposed to their own unregulated desires and ambitions, and they are exposed to profound changes in the organization of society. In particular, utilitarian individualism, which he thought was promoted primarily in the social thought of the English sociologist Herbert Spencer, encouraged egoism, ambition, and unlimited aspiration. The consequence of egoistic individualism (Marske 1987) is that the social malaise of a society without an adequate normative structure or 'anomie' is intensified, and in *Suicide* (Durkheim 1951) which he published in 1897, Durkheim attempted to show that the suicide rate was highest among those social groups which were most exposed to these anomic currents in society. Without normative restraint, individuals would succumb to such 'suicidal currents'.

In fact, Durkheim's argument in *Suicide* was far more complex than I have suggested, and he identified four different types of suicide, which have a specific causality. Some forms of suicide, such as fatalistic and altruistic suicide, are the products of too much regulation and social integration. Egoistic and anomic suicide were the types of suicidal behaviour which are most characteristic of contemporary society. Durkheim's analysis of suicide has been much debated and criticized (Atkinson 1978; Giddens 1965; Giddens 1966; Lukes 1973: 31), but I cannot in this introduction enter into this argument. The importance of *Suicide* for understanding *Civic Morals* is in terms of the light which it throws on Durkheim's critique of egoistic individualism as a process which uncouples the individual from the social structure.

Nisbet's third theme is the importance of authority in the conservative theory of society. The notion of authority 'runs like a leitmotif through all of Durkheim's works' (Nisbet 1965: 59). It is an essential feature of his view of morality, where authority, especially in the form of discipline, plays an important role in shaping 'personality' through moral education. Once more, Durkheim was particularly critical of the liberal utilitarian tradition of Bentham and James Mill, who, according to Durkheim, confused liberty with lawlessness. Without restraint and authority, human beings would be committed to a life of anarchy. The problem of modern society is indeed the slow erosion of moral authority, and the task of *Civic Morals* was to describe this crisis and to offer a set of solutions for the creation of authoritative moral guide-lines. The problem of modern society is to discover an effective principle which will give moral force and ethical authority to social norms and practices, without which discipline will be merely an external regulation. In *The Elementary Forms*, Durkheim wanted to show how obedience to religious practices produced self-restraint and altruistic actions produced personal asceticism as a necessary basis of social life as a whole. It is only on the basis of 'a certain disdain for suffering' (Durkheim 1961: 356), that society is possible at all.

This discussion allows Nisbet to get at the heart of Durkheim's conservatism, namely the centrality of religion, or more specifically the sacred, to Durkheim's sociological project as a whole. Here again Durkheim's approach departs significantly from the sociology of religion of Marx, Weber, or Simmel (Seger 1957; Turner 1991). Nineteenth-century theories of religion were largely individualistic and rationalistic, that is they treated religion as primarily a cognitive activity which was false from a scientific point of view (Goode 1951). Religion was the consequence of Man's misunderstanding of natural reality. For example, animism was an attempt to explain nature by reference to spirits. Since these theories are false from a positivistic perspective, religion will disappear with the advance of science. Durkheim departed radically from these cognitive orientations, by treating religion as social, collective, and practical. His theories of religion were heavily influenced by the arguments of William Robertson Smith whose *Lectures on the Religion of the Semites* (1889) showed how the sacrificial meal between men and the gods created a sacral community, and by Fustel de Coulanges's study of *The Ancient City* (1901) where the changing structure of classical society is examined in terms of theological changes.

In his religious studies, Durkheim attempted to show that Australian aboriginal totemism, as the simplest known religion, provided an insight into 'the elementary forms' of all religious life. His second task was to identify the genesis of the fundamental categories of human thought (such as time and space); this issue in the sociology of knowledge was also considered in *Primitive Classification* (Durkheim and Mauss 1963). His third objective was through an analysis of totemism to identify a number of generalizations about the universal functions of the sacred in social institutions.

Durkheim's work, which is a classic in the sociology of religion, has received ample commentary (Goode 1951; Pickering 1975;

Robertson 1970; Scharf 1970; Seger 1957; Turner 1991). The core of his argument proceeds along two lines. First, he attacked existing, typically individualistic, arguments about the nature of religion, in order to arrive at his own solution. For Durkheim, religion is a 'unified system of beliefs and practices relative to sacred things, that is to say, things set apart and forbidden – beliefs and practices which unite into one single moral community called a Church, all those who adhere to them' (Durkheim 1961: 62). His second line of approach was to argue that the 'elementary forms' of religion, by which he meant the basic structural characteristics of religion, provide an insight into social structures and processes as such. Religious beliefs are to be interpreted as the 'collective representations' of society; the unintended consequence of religious practices is to create a social bond; the practice of religious rituals creates a social enthusiasm or 'effervescence' by which social commitments are renewed; the training of the faithful in sacrifice and asceticism creates important norms of altruism and social service; and religious mythologies, which are dramatically re-enacted in the ritual, store up the collective memory of the social group, without which the continuity of this historical narrative of generations would be impossible (Wach 1944). Talcott Parsons was probably correct or at least insightful, when he argued that Durkheim, starting with the proposition that society is the basis of religion, concluded with the equally revolutionary equation that the basis of society is sacred. The problem of modern society is that we are in a transitional period; the old gods are dead, and new ones are yet to be born. Nationalism may prove to be such a god, inspiring devotion and sacrifice.

Finally, Nisbet argued that the underlying metaphor in Durkheim's sociology was that society is organic, and that its developmental laws can only be understood in terms of collective processes such as social differentiation which cannot be reduced to individual psychology, and especially to individual rationality. Against the utilitarian tradition, Durkheim rejected the idea that society was the result of a social contract drawn up between individuals, and that the development of society could be conceived in terms of an original contract (Abercrombie, Hill, and Turner 1986). Society is organic rather than contractual in Durkheim's more holistic perspective. He argued that a contract between individuals would be meaningless and ineffective unless it was based on deeply held values and beliefs, and unless it was sanctified by custom, ritual, and morality. The rejection of this utilitarian tradition occupied Durkheim in *The Division of Labor in Society* (1960), where he provided a specific attack on Spencerian sociology, but *Civic Morals* constitutes the core of Durkheim's critical offensive against individualistic/utilitarian accounts of property and contract; I shall turn shortly to this argument in detail in providing a description of the contents of his lectures on professional ethics and public morality. In conceiving of society as an organic whole and not as an aggregate of individuals, Durkheim has often been identified as a founder of 'structural-functionalism' as a distinctive school of sociology. Certainly Durkheim's view of historical change was primarily in terms of the dichotomy between mechanical and organic solidarity which he explored fully in *The Division of Labor in Society.*

This interpretation of Durkheim as a social theorist who laid the foundations for the analysis of social integration in social systems was promoted by Parsons in a number of major publications such as *The Social System* (Parsons 1991: 367ff), and in so doing Parsons has also, somewhat less directly than Nisbet, promoted the idea that Durkheim has to be seen as a theorist of social stability and social integration. For example, Parsons (1974) argued that Durkheim's account of solidarity in *The Division of Labor in Society* in terms of the *conscience collective* in mechanical solidarity in primitive societies and of social reciprocity in organic solidarity in advanced societies was a major solution to the Hobbesian problem of social order in the utilitarian tradition. Durkheim's analysis of the integrative functions of religious practice in both making and sustaining social communities provided Parsons with a theoretical source in classical sociology for his own emphasis on the importance of common values in the social cohesion of modern societies. In Parsons's early academic career, Weber's analysis of capitalism had been the primary intellectual stimulus for Parsonian sociology (Wearne 1989), but as Parsons moved more towards an analysis of the allocative and integrative requirements of a social system Durkheimian issues appear to have become increasingly important. Thus, Parsons's appreciation of the significance of the psychological internalization of values which he took from Cooley (Parsons 1968) was now supplemented by Durkheim's analysis of the integrative function of common beliefs to produce the cornerstone of Parsons's 'middle period', namely the internalization and socialization of values in social integration (Alexander 1984; Robertson and Turner 1989).

In the conventional paradigm of introductory textbooks for undergraduate sociology courses, there developed a tripartite version of classical sociology: Marx was a theorist of conflict and social change; Weber was a social philosopher of action and meaning; and Durkheim was a sociologist of social order, moral systems, and political stability. It has taken many years for a more complete interpretation of Durkheim to emerge, but recent perspectives on Durkheim have tended to take more notice of his political sociology (Giddens 1986; Lacroix 1981), his educational commentaries (Pickering 1979), the complexity of his methodological views (Gane 1988), his dependence on German moral philosophy (Meštrović 1991), his sociology of law and justice (Green 1989; Sirianni 1984), the richness of his views on cultural strains in advanced societies (Alexander 1988), and his awareness of the contradictions of modern society. These new emphases do not mean that previous perspectives on Durkheim were inaccurate or invalid; rather they produce an interpretation of Durkheim which is richer, deeper, and more comprehensive.

An alternative view of Durkheim's sociology was established by Alvin Gouldner's introduction (Gouldner 1962) to Durkheim's *Socialism*, which had been posthumously published in 1928. Generally speaking, Gouldner's aim was to show that Durkheim's sociology was the intellectual legacy of Henri Saint-Simon (1760–1825) rather than Auguste Comte (1798–1857). This interpretation was a subtle strategy to demonstrate Durkheim's link with socialism rather than conservatism. Gouldner shows that we can in fact read Durkheim's *The Division of Labor in Society* as a polemic against Comte. Durkheim's position was not that modern society cannot exist without consensus, but rather that the reciprocity of organic solidarity produces a basis of social order without a normative consensus. Second, Gouldner argued that anomie was not normlessness, but rather a disjuncture between existing norms and changing social structures. Third, the real dislocation of modern society was the absence of intervening social institutions between the individual and the state; occupational and professional associations were intended to fill this gap. Another dislocation of modern society was the division between local commitments to the nation-state and the growth of internationalism, cosmopolitanism, and globalism. It is true that Durkheim defines socialism as a moral regulation of the market place, but in *Socialism* he was concerned to understand how a moral regulation of the economy would be possible. Gouldner's work was thus important in reasserting the significance of Durkheim's interest in economic and political issues.

The trend in more recent interpretations, therefore, has been to assert that there was an important radical dimension to Durkheimian sociology which has been neglected as a consequence of the concentration on his arguments about social solidarity and his condemnation of economic individualism. Giddens has made the valuable point that Durkheim was not in fact strictly interested in '"order" in a generic sense, but of the form of authority appropriate to a modern industrial State' (Giddens 1986: 12). Furthermore, Durkheim's contributions to the sociology of law and the state were rather neglected by earlier interpretations of Durkheim (Pearce 1989); it is precisely in this context that we need to take his *Civic Morals* seriously, as Durkheim's most elaborate reflection on state power in relation to individual rights.

Durkheim's impact on France was in fact always regarded as dubious by the conservative wing of French politics, because Durkheimian sociology was identified with anti-clericalism and the Dreyfusard lobby (regardless of Durkheim's own views on these issues). Durkheim's Jewish background and his clear identification with 'the intellectuals' were sufficient to put him outside the conservative bloc in French society. We can identify one aspect of Durkheim's sociology which was especially critical of existing economic institutions, namely the inequality of wealth in France, which Durkheim regarded as particularly destabilizing. An important feature of his economic sociology was, thus, his bitter condemnation of the inheritance of property within a society which had an ideology of egalitarianism.

In terms of the actual interpretation of Durkheim's sociology, part of Nisbet's (1965) argument about the conservatism of Durkheimian thought was its dependence on the French tradition of conservatism which included de Maistre and de Bonald. More recent interpretations of Durkheim have identified his dependence on German philosophy, especially on Schopenhauer. The importance of this viewpoint is to place particular weight on Durkheim's sociology as a science of morality (Meštrović 1991). In this framework, we can see Durkheim's sociology as a reply to Kant's theory of morality and theory of knowledge. Briefly, we can see Kant's account of the moral imperative as an attempt to provide a rationalist justification for the Christian idea of brotherly (altruistic) love. Kant harnessed reason to ethics to explain why we should feel an obligation towards others; the categorical imperative claimed that we should treat others as we expected them to treat us. Morality was thus about reasonable obligation. Kant's epistemology and aesthetics ran in the same direction. Our knowledge of the world is determined or given by general categories of thought (cause, effect, time, and space). Knowledge is not imprinted on the mind by empirical reality. In the world of aesthetics, Kant argued that aesthetic judgement was disinterested, neutral, and objective; Kant thereby attempted to separate sensibility and aesthetics, because he denied that aesthetic judgements were emotive.

We can see Durkheim's account of ethics and knowledge as a reply to this Kantian legacy. In terms of the sociology of knowledge, Durkheim claimed, in *Primitive Classification* (Durkheim and Mauss 1963), that the fundamental categories of thought were located in the organization of society; social forms produced the forms of thought. For example, the analytical notion of 'space' is modelled on social space. In terms of religious belief, as we have seen, Durkheim derived the concept of 'god', or more exactly the dichotomy of sacred and profane, from social life; it is society which inspires in us the sense of the holy. In general, Durkheim wanted to deny that a rational appreciation of duty, or a utilitarian respect for sanctions, would ever be sufficient as a basis of moral commitment. Morality required compassion, fervour, and a sense of the sanctity of moral obligations to induce a sense of commitment and duty. In this respect, Durkheim followed Schopenhauer rather than Kant in formulating an empirical science of morals which would avoid the formal, a priori reasoning in Kantian moral philosophy.

Although Durkheim's sociology, such as his sociology of

education (Durkheim 1977; Pickering 1979), was clearly a social and political response to the crisis of French society in the late nineteenth century, his intellectual concerns were not unidimensionally driven by the legacy of French conservatism. While the intellectual legacy of Saint-Simon and Comte on Durkheim cannot be denied, Durkheim was also trying to come to terms with the intellectual legacy of Kant and Schopenhauer, and also with the impact in his own day of the political ideas of Heinrich von Treitschke whose pan-Germanism and state theory were condemned by Durkheim (Giddens 1986: 230) as dangerous doctrines.

One strong argument against the view that Durkheim was conservative can be taken from *Civic Morals* itself: namely in Durkheim's critique of the injustice which is associated with and an inevitable outcome of the institution of inheritance. Because wealth which results from inheritance has no necessary relationship to merit, Durkheim argued that it 'invalidates the whole contractual system at its very roots' (this vol.: 213). He then offered an attack on the social consequences of inheritance which would be worthy of Marx's prose:

Now inheritance as an institution results in men being born either rich or poor; that is to say, there are two main classes in society, linked by all sorts of intermediate classes: the one which in order to live has to make its services acceptable to the other at whatever cost; the other class which can do without these services because it can call on certain resources, which may, however, not be equal to the services rendered by those who have them to offer. Therefore as long as such sharp class differences exist in society, fairly effective palliatives may lessen the injustice of contracts; but in principle, the system operates in conditions which do not allow of justice.

(this vol.: 213)

Durkheim claimed that, with a growing sense of justice in a modern democracy, the institution of inheritance clashed with contemporary norms of equality. Unfortunately the English translation appears to be stale and cumbersome. The French 'la conscience morale', is rendered as 'men's conscience' and 'le sentiment' as 'attitude' with the result that Durkheim's text is psychologized and rendered sexist.

Durkheim proposed a moral principle of distribution which

would overcome these existing inequalities, namely 'the distribution of things amongst individuals can be just only if it be made relative to the social deserts of each one' (this vol.: 214). Now Durkheim's notion of justice is largely incompatible with a conservative theory of private property rights, in which the right of heads of households to dispose of their own property according to their own interests is a fundamental principle of 'possessive individualism' (Macpherson 1962).

Durkheim, however, recognized two forms of inheritance: wealth and talents. While the abolition of the privilege of inheritance would undermine economic inequality resulting from birth, the inheritance of talents, or what today we might call 'cultural capital' (Bourdieu and Passeron 1990), is equally significant and is not solely related to economic class. Thus *Civic Morals* concludes with the problem of the inheritance of talents:

To us it does not seem equitable that a man should be better treated as a social being because he was born of parentage that is rich or of high rank. But is it any more equitable that he should be better treated because he was born of a father of higher intelligence or in a more favourable moral *milieu*?

(this vol.: 220)

While Durkheim could see no ready social or political solution to this moral problem, he believed that only a special type of consciousness based on charity and human sympathy may overcome the tendency to judge the moral worth of a person in terms of their social background.

In a conclusion which goes back to Schopenhauer's idea that compassion is the root of moral action, Durkheim argued that charity 'ignores and denies any special merit in gifts or mental capacity acquired by heredity. This, then, is the very acme of justice' (this vol.: 220). These attitudes towards justice are hardly compatible with the social outlook of conservatism (Green 1989). In summary, an inspection of 'Durkheim's writings on the growth of moral individualism, on socialism, and on the State, in the context of the social and political issues which he saw as confronting the Third Republic, shows how mistaken it is to regard him as being primarily "conservative"' (Giddens 1986: 23). The political arguments of *Civic Morals* are particularly powerful evidence of such an interpretive 'mistake'.

INTELLECTUAL CONTINUITY

Civic Morals is also relevant to the debate about the thematic and intellectual continuity of Durkheim's sociology. Here again it was Parsons who, in *The Structure of Social Action*, had claimed that there was a profound discontinuity in the sociology of Durkheim. In particular, Parsons argued that Durkheim, especially in *The Division of Labor* of 1893 and *The Rules* of 1895, had embraced a positivistic theory of moral facts, while his later work such as *The Elementary Forms* of 1912 was based on idealism. Parsons attempted to show that a positivistic theory of morals which treats social facts as exterior, objective, and autonomous cannot solve the problem of how individuals become normatively committed to these moral facts. It fails to produce an adequate theory of internalization of moral facts, which can then become subjectively authoritative. It was only when Durkheim came to his final study of religion that he began to provide a theory of the emotive character of morality in terms of subjective affectivity. It is ritual practice and social effervescence which bring about the internalization of norms, but in Durkheim's sociology of religion these arguments are based on the view that 'society = god' – an equation which Parsons treated as idealism. In short, the analytical inadequacy of a positivistic theory of morals breaks down into idealism, producing an intellectual rupture at the heart of Durkheim's social science of morals. This position is difficult to sustain when we look at the development and contents of *Civic Morals*.

In many respects, Parsons's interpretation of Durkheim was a major intellectual advance at the time. His emphasis on Durkheim's critique of Spencer's individualism in Durkheim's development of the idea of the non-contractual element of contract in *The Division of Labor* was especially significant (Parsons 1981). However, a further feature of Parsons's interpretation involved dividing Durkheim's work into substantive topics. The 'early empirical work' concerned the occupational division of labour and the suicide problem in France, whereas 'the final phase' concerned religious ideas and questions of epistemology. The idea that Durkheim moved from empirical questions of social structure to epistemological problems of knowledge gave further credibility to the idea of a fissure in Durkheimian sociology. Again this division in terms of substantive areas is not tenable in terms of Durkheim's lectures on contract, property, the state, and religion in *Civic Morals*.

Professional Ethics and Civic Morals is in fact a collection of Durkheim's lectures, which had the title 'The nature of morals and of rights' and which Durkheim delivered at Bordeaux between 1890 and 1900. As Professor Kubali explains in his 'Preface' to the first edition of the lectures which Routledge & Kegan Paul published in English translation in 1957 (this vol.: ix–xi), these lectures existed in manuscript form with Marcel Mauss; three of the six lectures were published in the *Revue de Metaphysique et de Morale* in 1937, twenty years after Durkheim's death. It is unlikely that Parsons would have known of the existence of these lectures when he first published *The Structure of Social Action* in 1937. Parsons's bibliographical notes at the end of *The Structure of Social Action* contain some references to work from Durkheim's Bordeaux period, but there is no reference to the material which now constitutes *Civic Morals*. Durkheim's lectures appeared in French as *Leçons de Sociologie, physique des moeurs et du droit* (Durkheim 1950). The reader should note that, while Cornelia Brookfield's translation for Routledge & Kegan Paul in 1957 is technically accurate, it is often infelicitous and idiosyncratic.

These lectures on professional ethics, civic morals, property, and contract are thus from what Parsons regarded as Durkheim's 'positivist' phase; chronologically, they belong with *The Division of Labor* and *The Rules*. They were part of Durkheim's attempt to create an autonomous sociological discipline whose main subject matter was to be the study of moral facts. If we can show that the themes of *Civic Morals* in fact cover Durkheim's entire sociological interests, both 'late' and 'early', and if we can show that these interests were not bifurcated around the materialism/idealism dilemma, then we have shown that Durkheim's sociology was not ruptured by this dilemma. More importantly, we will establish an evolution of Durkheim's sociological ideas over a period of almost twenty years.

It is not necessary to provide a general summary of Durkheim's *Civic Morals*. George Davy's 'Introduction' to the 1957 edition (this vol.: xiii–xliv) remains an adequate general assessment and description. To avoid repetition, I shall focus on the core of Durkheim's argument across the six lectures with two questions in mind. First, do these lectures map out Durkheim's sociology as a whole by anticipating or discussing themes which are more fully developed later? Second, how does *Civic Morals* address theoretical or political issues which remain at the core of contemporary sociological debate? By answering these two questions, I hope also to address the query: why read Durkheim's *Civic Morals*?

The central issues of *Civic Morals* were part of the legacy from Saint-Simon in a number of important respects. The question behind Saint-Simon's work was the problem of the erosion of Christianity, the problems of industrialization, and the possibility of a 'religion of humanity' which, directed by sociological knowledge, would provide some coherence for a complex and differentiated society. Saint-Simon was thus exercised by the problems of political integration in Europe and, in various pamphlets and addresses, he attempted to conceptualize the institutions which would be necessary for a European parliamentary system (Taylor 1975).

A similar set of issues provided the structure of the argument in *Civic Morals*. Durkheim starts by providing an outline of the scientific study of morals, with which we are perfectly familiar from *The Rules* or *Suicide*, but much of the core argument of *Civic Morals* is focused on the problems of social integration and government authority in a post-Christian, if not secular, society in which there is a high degree of social differentiation. There is also concern for the problems of global social order and a form of government which might ultimately transcend the limitations of the nation-state.

The problem facing modern Europe is the separation of the economy from society and the absence of any effective regulation of the market place. The division of labour and the development of the modern economy have produced, or been accompanied by, the evolution of occupational groups and professional associations. For Durkheim, these groups offer some stability for modern society, but there has been no significant development of business professions; there is no code of conduct which can regulate economic activity. Professional associations are, in any case, rather local in their organization and social effects, but the crucial issue is the absence of a set of professional ethics. The crisis facing Europe is the anarchy of the market place and the underdevelopment of moral regulation. This 'moral vacuum' (this vol.: 12) can only be filled by the development of a 'corporate system', a code of business ethics, and state intervention in the market place. As Durkheim argued in *The Division of Labor* and *Suicide*, this lack of moral regulation means that individuals are exposed to the negative or anomic consequences of the business cycle and to their own unlimited desires and expectations. Durkheim was critical of this structural problem in modern societies, but he also attacked classical economists for failing to see the social consequences of unrestrained economic activity; economic functions were studied as if they had no social effects.

The social problem of modern society can be understood more effectively by examining the historical decline of the guild system which in Roman and medieval times had provided some ethical regulation of economic activity, by controlling their members, prices, and the conditions of exchange. Durkheim, in this historical sketch, adopts an argument which is characteristic of his sociological style of argumentation as a whole. He wants to show how these formal rules of conduct had a real effect on the behaviour of individuals, and he does this by a digression into religious history. The origins of the guild are to be found in the religious *collegium*. The cults which formed around craft activities provided festivals, feasts, collective sacrifices, and patterns of exchange such as the gift. These collective and ritual activities provided the social and moral force behind the original regulations on economic behaviour which are associated with the guild. In other words, Durkheim was never content to discover a formal obligation or regulation without showing how some collective and moral force brought about an affective commitment to some social practice. Morality and moral force is always to be discovered in 'something that goes beyond the individual, and to the interests of the group he belongs to' (this vol.: 24). This type of argument was put to special use in *The Elementary Forms*, but we can also see it in operation in *Civic Morals*.

The crisis in the European socio-economic system can be best resolved through the development of a corporative structure which would organize the various branches of industry, provide an administrative and electoral system by which interests could be articulated, and thereby come to develop a macro-system of moral authority and regulation. The 'cells', so to speak, of this structure would be the professional and occupational associations to which individuals would be attached. However, the whole system could only function if the state became 'the central organ' of the whole system (this vol.: 39). These relations between the individual, the professional associations, and the state are the subject matter of civic morals. These civic morals thereby determine the normative relationships between the individual and the state.

Given the widely held opinion that Durkheim failed to develop an adequate political sociology, it is a striking feature of *Civic Morals* that Durkheim gives special prominence to the state in regulating society and directing moral activity. For Durkheim, the state has the responsibility to 'work out certain representations which hold good for the collectivity ... [because] the State is the very organ of social thought' (this vol.: 50–1). We can only understand these arguments by realizing that here, as elsewhere in Durkheim's sociology for example in *The Division of Labor*, he is attempting to counteract the arguments of Herbert Spencer. For Durkheim, the contemporary development of the state is not incompatible with the growing importance of the individual and individualism. The state is an essential feature of the evolution of individual rights, because it is only the state which has sufficient authority and collective power to create and protect individual rights. There is nothing about the state which must produce a political tyranny, and indeed it is the modern state which has liberated the individual from the particularistic forms of domination, which were typical of feudalism.

Of course, while Durkheim overtly developed a political sociology in *Civic Morals*, his political ideas assumed a moral framework (Wallwork 1972: 103). The role of the state 'is to persevere in calling the individual to a moral way of life' (this vol.: 69) and this leads Durkheim to the assertion that the state is 'the organ of moral discipline' (this vol.: 72). One of the few recent studies in historical sociology which has recognized the importance of Durkheim's political ideas in *Civic Morals* is *The Great Arch* (Corrigan and Sayer 1985: 6), but Durkheim is criticized for not recognizing that the moral authority of the state results from a social struggle over morality, and that moral regulation has to be enforced. In general terms, one objection to Durkheim's account of the state as a moral agency is the optimistic belief that state terror is not a significant problem. In a valuable analysis of Durkheim's sociology, Edward Tiryakian has suggested that Durkheim's optimistic view of the state as an institution which protects the individual from particularistic patterns of oppression is a consequence of Durkheim's Jewish origins. The fact that the First Republic in France had emancipated the Jews from the restrictions of the *ancien regime* provided Durkheim with a concrete model of the state as the protector of individual freedoms (Tiryakian 1978: 198).

There is, however, an important additional element to Durkheim's argument which we should not ignore. For Durkheim, the possibility of state tyranny is limited by the presence of intermediary institutions between the individual and the state. In this respect, Durkheim's argument followed closely on Alexis de Tocqueville's analysis in 1832 of the problems of democracy in America, where he had claimed that a democratic despotism could only be avoided if there was an effective system of voluntary associations acting as a social buffer against political domination. This question of intermediary groups was essential to Durkheim's treatment of democracy in Chapter 9. For Durkheim, the 'political malaise' has the same origin as the 'social malaise', namely 'the lack of secondary cadres to interpose between the individual and the State ... these secondary groups are essential if the State is not to oppress the individual: they are also necessary if the State is to be free of the individual' (this vol.: 96). These 'cadres secondaires intercales entre l'individu et l'Etat' (Durkheim 1950: 116) are essential both to individual liberties and to the effectivity of the state. Here again Durkheim looked towards professional and occupational associations to form this necessary intermediate stratum of institutions between the individual and the state. Indeed, these professional associations would become, according to Durkheim, the very foundation of political life. These institutions were necessary to avoid what we might usefully call 'political anomie'.

With these institutions and associations, the danger of political despotism would recede, but this situation still leaves open the problem of political commitment. What is the root of loyalty to the state? In Durkheim's argument in *Civic Morals*, to which he returned during the First World War, it is patriotism which is the core of this political commitment. Once more we can identify a typical Durkheimian sociological argument in his treatment of political loyalties. Patriotism is a type of secular religion, and thus it is possible to talk about a 'cult of the state' in which citizens are, as it were, the worshippers. Patriotism is constituted by 'the ideas and feelings as a whole which bind the individual to a certain State' (this vol.: 73). But Durkheim thought that in modern times, there were two forms of political loyalty which he called 'patriotism' and 'world patriotism'. For some reason, Cornelia Brookfield has translated 'le cosmopolitisme' as 'world patriotism' which does not adequately convey Durkheim's meaning (Durkheim 1950: 87). In ancient times, this division did not exist, because only one cult was possible: 'this was the cult of the State, whose public religion was but the symbolic form of the State' (this vol.: 72). The evolution of modern society has produced a wider horizon for human consciousness as human beings become conscious of their involvement in 'humanity' on a global basis. Consciousness becomes more universal under these new conditions. This universal consciousness is at a higher moral level than mere patriotism, and the importance of this emerging universal consciousness is that it becomes possible 'to imagine humanity in its entirety organized as a society' (this vol.: 74). Thus, in an argument which was very close to Saint-Simon's vision, Durkheim anticipated the idea of political globalization on the basis of a universalistic notion of humanity (Turner 1990). This idea of cosmopolitanism was part of Durkheim's later condemnation of pan-Germanism and a feature of his critique of war 'which reduces societies, even the most cultivated, to a moral condition that recalls that of the lower societies. The individual is obscured' (this vol.: 117). The actual expression used by Durkheim was much stronger: 'L'individu disparaît' (Durkheim 1950: 140).

This brief commentary on some of the key ideas of *Civic Morals* is intended to argue that Durkheim's sociology is not in any simple sense 'conservative' and to show that there was no significant fissure or rupture in the development of his ideas. The Bordeaux lectures contained the essential features of his sociology as a whole, both in its 'early' and 'late' topics of analysis. One final argument which can support this claim is to note the importance of Durkheim's criticisms of Kantian philosophy in the final chapters on property.

I have already noted that in general we can see Durkheim's sociology as a response to Kant. First, in his sociology of knowledge, for example in *Primitive Classification* (Durkheim and Mauss 1963), Durkheim argued that our knowledge of the world was not grounded in a set of a priori categories which were the structure of (the individual) mind. The fundamental categories of thought (time, space, causation and number) were collective and social categories which were modelled on the structure of society itself. The fundamental categories of mind are in fact collective representations of social life. This theory has, of course, been challenged on philosophical grounds (Needham 1963), but I am only concerned at this stage to show the Kantian origins of this debate in Durkheim. In a similar fashion, Meštrović (1991) has convincingly demonstrated that Durkheim's moral and aesthetic views were an attempt to criticize the individualism and rationalism of Kant's moral argument concerning the nature of the categorical imperative. This debate with Kant was equally important in *Civic Morals*.

In a complex argument about the collective and sacred origins of property, in which ownership is compared with the concept of the taboo, Durkheim attempted to present a sociological alternative to Kant's discussion of property. After a lengthy and difficult argument, which the reader must study in detail, Durkheim came to a striking conclusion. Respect for property is not related to the individual personality; this respect has an origin which is exterior to the individual; it is, once more for Durkheim, an issue concerning the sacred/profane dichotomy. Thus 'Property is property only if it is respected, that is to say, held sacred' (this vol.: 159).

Where does the sense of the sacred come from? Because the arguments of *The Elementary Forms* are probably quite familiar to the sociologist, we can easily anticipate how Durkheim will answer this question. However, from the point of view of exegesis, it is interesting to see how fully *Civic Morals* rehearses the arguments of his 'final phase'. For Durkheim, we can best understand religion as the 'way in which societies become conscious of themselves and their history' (Durkheim 1957: 160). In a thesis which reproduced Fustel de Coulanges's ideas about the ancient world, Durkheim argued that

The gods are no other than collective forces personified and hypostasized in material form. Ultimately, it is the society that is worshipped by
the believers; the superiority of the gods over men is that of the group
over its members. The early gods were the substantive objects which
served as symbols to the collectivity and for this reason became the
representations of it.

(this vol.: 161)

Once more the French original is inevitably superior:

Les dieux ne sont autre chose que des forces collectives incarnées,
hypostasiées sous forme matérielle. Au fond, c'est la société que les
fidèles adorent; la supériorité des dieux sur les hommes, c'est celle du
groupe sur ses membres. Les premiers dieux ont été les objets matériels
qui servaient d'emblèmes à la collectivité et qui, pour cette raison, en
sont devenus les représentations.

(Durkheim 1950: 190–1)

This argument, although it was anticipated by Fustel de
Coulanges and William Robertson Smith, was the decisive origin
of a *sociological* theory of religion which attempted to locate the
nature of religious belief and experience in collective life. The
experience of the holy is, in Durkheim's argument, produced by
an exterior and superior authority (the society) through collective
rituals, which in turn create a social effervescence. It was on this
basis that Durkheim challenged the individualistic rationalism,
not only of Kant's moral philosophy, but also of Kant's Protestant
version of faith.

Thus, in its account of the nature of moral facts, the importance of civic morals and professional associations in regulating
the relationship between the individual and the state, in its
critique of economic individualism in Spencerian sociology and
English utilitarianism, in its analysis of the social and political
malaise of modern society, in its analysis of the social origins of
religious ideas, and in its general critique of Kantian philosophy,
Civic Morals provides a summary of Durkheim's sociology as a
whole. That this volume should accomplish such a synoptic task
is not surprising. Although these lectures on morals and rights
were given originally in Bordeaux in the 1890s, they were
repeated in 1904 and 1912 in the Sorbonne. In Parsons's exegetical framework these lectures therefore cover both
Durkheim's 'early empirical work' and the 'final phase'. In the

light of Durkheim's concern in the period 1912 to 1917 with patriotism, war, and religion, the continuities between his work at Bordeaux in the 1890s and later at the Sorbonne are more impressive and obvious than the alleged discontinuities. The central element of this continuity is that what we common-sensically think of in individual terms (mind, feelings, commit-ment, and so forth) are collective and social. In more specific terms, the force of emotive commitment to moral rules has to be found in compassion, and compassion has its origin in the sacred/collective character of social life.

CONCLUSION

In a recent collection of essays on *Durkheimian Sociology* (Alexander 1988), Randall Collins has noted that 'Of the great classic figures of sociology, at the present time Durkheim's repu-tation is at its lowest' (Collins 1988: 107). This low ebb is partly explained by the ways in which Durkheim has been (mistakenly) interpreted and received. Durkheim's reception has been initially through the English functional anthropologists and later he was embraced as the founder of multivariate statistics on the basis of his arguments in *Suicide*. Durkheim was also appropriated by conservatism as a leading figure. For Collins, Durkheim's work should be seen as a valuable contribution to conflict analysis, because, on the basis of *The Elementary Forms*, he produced a powerful theory of social solidarity, and this concept is essential to the development of conflict sociology. The abiding problem of Durkheimian sociology, however, is that it 'tended to minimize the significance of social classes and their conflicts' (Collins 1988: 109). On the basis of a close inspection of *Civic Morals*, we can see that Collins's judgement is erroneous.

Civic Morals is dominated by a political and social analysis of the malaise of modern society, which is the failure of inter-mediary institutions to provide a linkage between the state and the individual. Professional ethics and professional associations, along with the development of a system of civic morals, are seen to be an antidote to this problem. However, Durkheim also believed that the state would have to help create a corporative system, in which an organic bond would emerge between the individual and the state. Without these associational and legis-lative changes, the anarchy of the economy would continue and society would remain anomic. However, despite these changes, the fundamental inequality between social classes would remain, because the system of inheritance guaranteed the unequal distribution of property in society across generations. This property system would continue to destabilize and delegitimize society, regardless of changes in the state and the professions. Durkheim's sense of justice was outraged by this inequality, because economic inequality prevented the development of compassion which he thought was the foundation of morality.

In retrospect, Gouldner's introduction to Durkheim's *Socialism* still provides us with one of the most accurate insights into the real nature and purpose of Durkheimian sociology. Gouldner argued correctly that one of the central issues in *Civic Morals* concerned inheritance and its moral consequences. Thus, Durkheim 'holds that it is the existence of social classes, characterized by significant economic inequalities, that makes it in principle impossible for "just" contracts to be negotiated' (Gouldner 1962: 30). *Civic Morals* can thus be read as a treatise on the problem of justice in modern societies, and how the sense of injustice in relationship to the property system is a feature of their political instability. Durkheim's attempted answer to this problem was socialist in arguing that a new economic order – a corporative system – would be required to function as a replacement for the archaic order of the guild.

Durkheim's sociology of morals was not, therefore, a conservative theory of social order. It was a political response to the malaise which he saw in France in the second half of the nineteenth century, but it was also a socialist response to the negative impact of an anarchic economy on moral life. The capitalist economy had become differentiated from political and social life. In the absence of a system of moral regulation, capitalist economic relations would be regarded as illegitimate, because capitalist forms of inheritance and property relationships failed to reward merit and effort adequately. Durkheim did not look back with nostalgia to the medieval guild system, because he realized that the guild was no longer adequate to the task. The reform of an anomic society required a radical approach to political change and economic organization. Durkheim's analysis of the instabilities of a society based upon contract, his defence of the state as the basis of individual rights, and his ethical critique of inequality and inherited wealth have a clear relevance to the social and political problems of the late twentieth century. This relevance is one important reason for reading *Professional Ethics and Civic Morals*.

REFERENCES

Abercrombie, N., Hill, S., and Turner, B.S. (1986) *Sovereign Individuals of Capitalism*, London: Allen & Unwin.

Alexander, J.C. (1984) *Theoretical Logic in Sociology, The Modern Reconstruction of Classical Thought, Talcott Parsons*, London: Routledge & Kegan Paul.

Alexander, J.C. (ed.) (1988) *Durkheimian Sociology: Cultural Studies*, Cambridge: Cambridge University Press.

Atkinson, M. (1978) *Discovering Suicide*, London: Macmillan.

Black, A. (1984) *Guilds and Civil Society in European Political Thought from the Twelfth Century to the Present*, London: Methuen.

Bourdieu, P. and Passeron, J.-C. (1990) *Reproduction in Education, Society and Culture*, London: Sage.

Cobban, A. (1957) *A History of Modern France*, Harmondsworth: Penguin Books, 2 vols.

Collins, R. (1988) 'The Durkheimian tradition in conflict sociology', pp. 107–128 in J.C. Alexander (ed.) *Durkheimian Sociology*, Cambridge: Cambridge University Press.

Corrigan, P. and Sayer, D. (1985) *The Great Arch*, Oxford: Basil Blackwell.

Coulanges, F. de (1901) *The Ancient City*, Paris.

Durkheim, E. (1915a) *Qui a voulu la guerre?*, Paris: Colin.

Durkheim, E. (1915b) *L'Allemagne au-dessus de tout*, Paris: Colin.

Durkheim, E. (1950) *Leçons de Sociologie, physique des moeurs et du droit*, Paris: Presses Universitaires de France.

Durkheim, E. (1951) *Suicide, A Study in Sociology*, New York: Free Press.

Durkheim, E. (1960) *The Division of Labor in Society*, New York: Free Press.

Durkheim, E. (1961) *The Elementary Forms of the Religious Life*, New York: Collier Books.

Durkheim, E. (1962) *Socialism*, New York: Collier Books.

Durkheim, E. (1964) *The Rules of Sociological Method*, New York: Free Press.

Durkheim, E (1969) 'Individualism and the intellectuals,' *Political Studies* 17: 19–30.

Durkheim, E. (1977) *The Evolution of Educational Thought*, London and Boston: Routledge & Kegan Paul.

Durkheim, E. and Mauss, M. (1963) *Primitive Classification*, London: Cohen & West.

Fenton, S. (1984) *Durkheim and Modern Sociology*, Cambridge: Cambridge University Press.

Feuer, L.S. (ed.) (1969) *Marx and Engels, Basic Writings on Politics and Philosophy*, London: Fontana.

Freud, S. (1930) *Civilization and its Discontents*, London: Hogarth Press.

Gane, M. (1988) *On Durkheim's Rules of Sociological Method*, London and New York: Routledge.

Giddens, A. (1965) 'The suicide problem in French sociology', *British Journal of Sociology* 16: 3–18.

Giddens, A. (1966) 'A typology of suicide', *European Journal of Sociology* 7: 276–95.

Giddens, A. (1978) *Durkheim*, London: Fontana.

Giddens, A. (ed.) (1986) *Durkheim on Politics and the State*, Cambridge: Polity.

Goode, W.J. (1951) *Religion among the Primitives*, Glencoe, IL: Free Press.

Gouldner, A.W. (1962) 'Introduction' to Emile Durkheim, *Socialism*, New York: Collier Books.

Green, S.J.D. (1989) 'Emile Durkheim on human talents and two traditions of social justice', *British Journal of Sociology* 40: 97–117.

Lacroix, B. (1981) *Durkheim et le politique*, Montreal: Presses de l'Université de Montreal.

Lukes, S. (1973) *Emile Durkheim, His Life and Work, A Historical and Critical Study*, London: Allen Lane.

Macpherson, C.B. (1962) *The Political Theory of Possessive Individualism*, Oxford: Clarendon.

Marske, C.E. (1987) 'Durkheim's "cult of the individual" and the moral reconstitution of society', *Sociological Theory* 5: 1–14.

Meštrović, S.G. (1991) *The Coming Fin de Siecle, An Application of Durkheim's Sociology to Modernity and Postmodernity*, London and New York: Routledge.

Miquel, P. (1968) *L'Affaire Dreyfus*, Paris: Presses Universitaires de France.

Nandan, Y. (ed.) (1980) *Emile Durkheim: Contributions to L'Année Sociologique*, New York: Free Press.

Needham, R. (1963) 'Introduction' to E. Durkheim and M. Mauss, *Primitive Classification*, London: Cohen & West.

Nisbet, R. (ed.) (1965) *Emile Durkheim*, Englewood Cliffs, NJ: Prentice-Hall.

Nisbet, R. (1967) *The Sociological Tradition*, London: Heinemann Educational Books.

Orru, M. (1987) *Anomie: History and Meanings*, London: Allen & Unwin.

Parsons, T. (1937) *The Structure of Social Action*, New York: McGraw-Hill.

Parsons, T. (1968) 'Cooley and the problem of internalization', in Albert J. Reiss Jr (ed.) *Cooley and Sociological Analysis*, Ann Arbor: University of Michigan Press.

Parsons, T. (1974) 'Introduction' to Emile Durkheim, *Sociology and Philosophy*, New York: Free Press.

Parsons, T. (1981) 'Revisiting the classics throughout a long career', pp. 183–94 in Buford Rhea (ed.) *The Future of the Sociological Classics*, London: Allen & Unwin.

Parsons, T. (1991) *The Social System*, London: Routledge.

Pearce, F. (1989) *The Radical Durkheim*, London: Unwin Hyman.

Pickering, W.S.F. (1975) *Durkheim on Religion: A Selection of Readings with Bibliographies*, London: Routledge & Kegan Paul.

Pickering, W.S.F. (ed.) (1979) *Durkheim, Essays on Morals and Education*, London: Routledge.

Robertson, R. (1970) *The Sociological Interpretation of Religion*, Oxford: Basil Blackwell.

Robertson, R. and Turner, B.S. (1989) 'Talcott Parsons and modern social theory: an appreciation', *Theory Culture & Society* 6: 539–58.

Scharf, B.R. (1970) 'Durkheimian and Freudian theories of religion: the case for Judaism', *British Journal of Sociology* 21: 151–63.

Seger, I. (1957) *Durkheim and his Critics on the Sociology of Religion*, Columbia University, Monograph Series, Bureau of Applied Social Research.

Sirianni, C.J. (1984) 'Justice and the division of labour: a reconsideration,' *Sociological Review* 32: 449–70.

Smith, W.R. (1889) *Lectures on the Religion of the Semites*, Edinburgh.

Stauth, G. and Turner, B.S. (1988) *Nietzsche's Dance, Resentment, Reciprocity and Resistance in Social Life*, Oxford: Basil Blackwell.

Taylor, K. (1975) *Henri Saint-Simon 1760–1825, Selected Writings on Science Industry and Social Organisation*, London: Croom Helm.

Tiryakian, E.A. (1978) 'Emile Durkheim', pp. 237–86 in Tom Bottomore and Robert Nisbet (eds) *A History of Sociological Analysis*, London: Heinemann.

Tönnies, F. (1957) *Community and Association*, Michigan: Michigan State University.

Turner, B.S. (1990) 'The two faces of sociology: global or national', *Theory, Culture & Society* 7: 343–58.

Turner, B.S. (1991) *Religion and Social Theory*, London: Sage.

Wach, J. (1944) *Sociology of Religion*, Chicago and London: University of Chicago Press.

Waldron, J. (ed.) (1987) *Nonsense upon Stilts, Bentham, Burke and Marx on the Rights of Man*, London and New York: Methuen.

Wallwork, E. (1972) *Durkheim, Morality and Milieu*, Cambridge, MA: Harvard University Press.

Wearne, B.C. (1989) *The Theory and Scholarship of Talcott Parsons*, Cambridge: Cambridge University Press.

INTRODUCTION

BY GEORGES DAVY

Doyen de la Faculté des Lettres Université de Paris

SOME aid may be needed to understand these unpublished lectures of Durkheim and what he meant by the nature of morals, and why, in a study of ethics, he gave priority to a description of morals and in sociology in general, to the definition and observation of facts. I want therefore to bring out the main themes of the doctrine and the chief precepts of the method of a man who is recognized as the founder of French sociology.

At the outset, we see there are two themes of equal importance. We have to begin by separating them, to see how they diverge. Then we have to bring them together to understand how they are reconciled and how they give to sociology its starting-point and set the direction for its development. One is the theme of the science and the other that of society or 'the social'. The first is concerned with the mechanism and is quantitative, and the other with what is specific and qualitative.

No one could have any hesitation in naming the theme of society or sociality as the primary one, if he opens that standard work of the sociologist, the little book that appeared in 1895 with the title *Les Règles de la Méthode Sociologique*. His eye will at once light on the first chapter—"What is a social fact?". He will also, without surprise, see first place given to defining the subject of the new study, the social fact—laid down as specific and irreducible to any single element that contains the germ of it. Does not this fact, considered from the precisely social angle, correspond exactly to the term sociology, and at the same time provide its subject? We have mentioned the theme of science first, not because we fail to give 'the social' its due but because this theme of science throws light on the primary aim of the doctrine and outlines the nature of the method.

xliii

First, the aim : and let us say more comprehensively, the aim and the occasion. It is true that neither one nor the other is anything new. On the contrary, both link the author with a philosophic ancestry—recently, with Auguste Comte and Saint-Simon—and—remotely, with Plato. Plato's philosophy did not exclude politics any more than morals. For him, the two subjects 'Of the State' and 'Of Justice' were the same, and he dreamed of protecting the City from disorder and excess by means of the wisest possible constitution: this he conceived to be founded only upon science—not on opinion alone—upon a science which no doubt to him was not yet the science of facts, as it was to be in the positivist sociology of the nineteenth century. This science of ideas, as he conceived it, was none the less in his eyes the only true science and the only salvation for man and for the City. Auguste Comte, nearer to our own time, was confronted with a similar occasion of political and moral crisis, in his particular case provoked by the French Revolution and by the reconstruction that had to follow so much thrown over; he sought from science, but in a positive sense, the secret of the mental and moral regeneration of humanity. And it is this same salvation through science that was sought for so passionately by Durkheim, after he had seen the convulsion of minds and institutions in France that followed on the defeat of 1870 and witnessed the shock of another kind, with a similar call for reconstruction—the upheaval caused by the rise of industry. The vast changes in things called for reforms in the affairs of men. It must be the function of science alone to inspire, to direct and carry out the reconstruction needed and, since the crisis is one of societies, the science to resolve it must be one relating to societies, or social science. Such is the conviction out of which the sociology of Durkheim springs. It is also one on which his sociology stands—a system born out of that same absolute faith in science as the politics of Plato or the positivism of Auguste Comte.

We shall go on to describe how this science of societies is at the same time the science of man and to what degree it is so. Further, we shall relate how the knowledge of the nature of man (always the true goal of philosophy from its earliest days) seeks to raise itself, with the humane sciences, to the same level of objectivity as that of the sciences proper. It is, however,

to social science or sociology in its *exact sense*, that we shall at
the outset apply this objectivity that Durkheim (without any
real reason perhaps) would refuse to extend to every aspect
of man. He would confine it to one aspect alone, the one we
propose to call his social dimension. This, it is true, is only a part
of the human sum, but in the eyes of the author it is the only
part—excluding the individual—that yields to expounding by
science.

Hence, we have the theme of science predominating in the
execution as in the initial intention. But to make it possible to
deal with society in a scientific way, it still remains for society
to present a true reality to science, a datum which should be
the proper subject of social science. And here appears, equal in
importance with the other theme and bound up with it, the
theme of 'the social' (as defined in Ch. I of the *Règles*, referred
to above), to establish the specific nature of this subject.
'The social' is to be recognized by certain signs: by the *exteriority* it exhibits and by the *constraint* it exercises over the
individual. Its true essence, however, lies beyond these signs,
in the fact originating (to the extent of being necessary to it)
from the grouping as such, and especially from the human
grouping.

It has indeed been found possible to describe animal societies,
but without success in finding within them the secret of human
societies, despite undoubted analogies. Thus, not deduction
but only comparison is to be drawn from biology, which gives
sociology no more than its foundation. The only societies
properly so called are those of man and of this Durkheim was
convinced. This not only confirms the specific nature of 'the
social', to which he held so firmly, but makes of social science
essentially a human science: society is a human adventure.
Thus, the fundamental fact of the grouping must be understood
as coming within the human order. It is there that we discern
the character of the grouping as a phenomenon—a character
that at once unifies, constructs and gives meaning.

This is, it follows, its primary character and one which does
not allow of its being relegated to anything more elementary
or fundamental than itself. Whilst it is conceded that the fact of
the grouping is not posterior to the existence of the individual,
it is equally valid that it is not anterior. That is so because

individuals would not exist but for the grouping and the grouping has no existence without the individuals. A society in a state of vacuity is as much a figment of the imagination as an individual who is absolutely solitary and without place in any society. Individuals must be conceived as component parts of an organism. It is likewise from their whole community that they derive their regulation, their position and finally their existence, which must be qualified as a 'being-within-the-group'. The humanity of man is only conceivable within the human aggregation and, in one sense at least, as existing through it.

To affirm the specific reality of 'the social' thus consolidates the social whole with its parts. It in no way gives it separate, distinct form outside of them, as one might have been led to believe by the attributes of exteriority and constraint, into which we have often been tempted to read more than mere signs. We know how Durkheim, in this connection (both in the introduction to the 2nd Edition of the *Règles* and on many other occasions), disclaimed having gone back on his theory of positivity and having given life to a mere fiction. Nor, when 'the social' takes on the features of the collective consciousness, would he accept it as anything other than associated consciousnesses and the framework by which consciousnesses are associated.

We do not depend on a study of the famous article on individual representations and collective representations to perceive that although the analysis of the social fact sometimes puts a strain on the language in order to emphasize its objective reality, it does not therefore exclude all psychological components.

La Division du Travail (2nd Ed., p. 110) already recognizes that social facts are produced by an elaboration *sui generis* of psychological facts, an elaboration having some analogy with that produced in each individual consciousness and which "transforms progressively the primary element (sensations, reflexes, instincts), of which it is originally composed". On the subject of the collective consciousness (to which crime is an offence as an attack on its very existence, calling for retribution), do we not find in the same book (p. 67) the following psychological analysis: "This representation (of a force which

we feel vaguely as being outside and above us) is certainly
illusory. It is in ourselves and in our selves alone that the
offended sentiments are to be found. But this illusion is necessary. Since, as a result of their collective origin, their universal-
ity, their permanence in time and their intrinsic strength,
these sentiments have an exceptional force, *they are radically
separate from the rest of our consciousness* (our italics), whose states
of being (*états*) are far weaker. They dominate us. They have
something of the superhuman about them, as it were; and at
the same time they attach us to objects which are outside our
temporal life. Thus they appear to us as the echo within us of a
force which is extraneous to us and, what is more, superior to
the force we are. We are thus compelled to project the sentiments beyond ourselves and to relate what concerns them to
some exterior object." The author goes so far as to speak in this
connexion of partial derangements in the personality, of some
inevitable mirage. Following this, the conclusion of his analysis
comes back from the psychological to the sociological. " Besides ", he writes in effect, "the error is only partial. Since
these sentiments are collective, it is not ourselves that they
represent in us, but 'the society '." Of the collective consciousness as constituted he has to say (*ibid.* p. 46): "There is no
doubt that the sub-stratum does not consist of one sole organ.
It is diffused, by definition, through the whole range of society.
But none the less, it possesses specific features which make a
separate reality of it. It is indeed independent of the particular
conditions in which individuals are placed: they pass on and it
remains. . . . It is therefore something quite different from the
individual consciousnesses, although it only becomes reality
through the individuals. It is the psychological type of society,
the type which has its own properties, its conditions of existence,
its own way of development, exactly as the individual types
have, but in a different way." As we see, we are far indeed
from the so-called definition of the social phenomenon which
would make of collective consciousness a thing *per se*. Here we
have, on the contrary, the Durkheim definition opening out on
to a whole social psychology that we see mapped out both
in the important preface to the *Règles* as re-issued and in the
article quoted above on collective representations.

Such, then, is the kind of reality which it is proper to assign to what is called the social fact or collective consciousness: an absolute and over-riding fact relating to the group; an echo in the consciousnesses but one which is heard only in the group consciousnesses. It is an immanence going always from the whole to each of the parts and one that takes on an aspect of transcendancy only by projection. This comes about as a result of the more or less conscious sentiment or awareness that each component part has (by its very participation in its whole) of being drawn out of its passivity (which can only go on repeating itself without end), and being called upon to play a proper part in the common harmony, a part which gets its bearings from the superior unity of the whole.

But although 'the social' may indeed have this reality of its own, as just defined, which neither biology nor psychology, by breaking up its complex unity, can deprive it of; and although, therefore, sociology is not lacking in a purpose, it must also—if it is to be a science—not be lacking in objectivity. And here the theme of science recurs (that we held to be indivisible from that of sociality) which, for it to be rightly a science, imposes this precept upon sociology—that is, to treat social phenomena as things. Here again, we must avoid ambiguity in the word 'thing'. It is not a matter of seeing only a material datum in the social phenomenon. Durkheim always disclaimed any such materialism—but he was only against regarding it as a given fact—*given*, that is, like any thing we encounter just as it is and not imagined or fabricated according to what one believes it might be or desires it to be. That said, the fact being given as a thing in no way presumes that it should be only a material *thing*, and in no way excludes its also, or at the same time, being an *idea*, a belief, a sentiment, a habit or behaviour, which are, no less than matter, realities, existing and having effect and therefore capable of being observed.

Now it is precisely this observability that we want to underline when, in speaking of 'the social', we put forward the exteriority which is given as its symbol. And so that this possibility of objective observation shall not escape us or be compromised, Durkheim proposes to tackle 'the social', anyway at first, in its most exterior aspect, the symbol perhaps of a conscience not directly accessible, but in any case a reality which does not elude observation. This reality has only one constituent; and is collective, and thus involves considerable and repeated manifestations—tempting subjects for comparisons and statistics. The same reality is moreover an institution, this time crystallized in political forms or in codes and rituals, that is, changed into things that are easily observed. This is the route Durkheim takes in his *Division du Travail Social:* by a method strictly analogous to that of the psychology of behaviour, he seeks to discern social solidarity through its various forms; this he does through its observable manifestations —(sanctions of restrictive or restitutional law), or through the behaviour which it inspires—(co-operation or working for the common good). He goes to work along the same lines in yet another of his writings where, by reference to suicide and homicide rates as shown in statistics, he attempts to estimate attachment to life, respect for the person and the need of integration, during a given period, in a given society or class.

This starting point of the method is of such importance that we should let the author speak for himself: "In submitting an order of facts to scientific enquiry," he declares, "it is not enough to observe them carefully, to describe and classify them, but what is far more difficult, one must further, to use the words of Descartes, get the angle (*le biais*) by which they become scientific; that is, discover in them some objective element which allows of precise determination and, if possible, of their measurement. We have tried to fulfil this condition demanded by every science. It must be very clear that we have studied social solidarity through the system of juridical rules; how, in the search for causes, we have set aside all that might be too much affected by a personal view or by subjective appraisal, so that we might get at certain facts of social structure that lie deep enough to qualify as objects of comprehension and hence, of science." (*Div. du Travail*, Pref. p. XLII) A few pages further on, we read more explicitly still: "Social solidarity is an entirely moral phenomenon which in itself does not yield to precise observation or, above all, to any measurement. To arrive at this classification and this comparison, we must substitute for the inward fact that escapes us an exterior fact that stands as a symbol for it, and study the first by way of the second. This visible symbol is the law. In fact, wherever there is social solidarity, in spite of its immaterial nature it does not remain in a state of pure potentiality but manifests its presence by effects perceptible to the senses. The more solidarity there is amongst the members of a society, the more they maintain various relations, whether it be one with another, or with the group collectively. For if they seldom came together, they would depend on one another only in a tenuous and fitful way. These relations, on the other hand, are inevitably in proportion to the juridical rules that determine them."

"Indeed, social life, wherever it exists in any permanent way inevitably tends to assume a definite form and to be organized: the law is precisely this very organization in its most stable and clearly-defined form. The life in general of the society cannot extend at any point without its juridical life extending at the same time and in the same ratio. We may therefore be certain of finding all the essential varieties of social solidarity reflected in the law." (*Div. du Travail*, pp. 28–9). This leads to the conclusion:—"Our method therefore lies mapped out. Since we find the main forms of social solidarity reproduced in the law, we have only to classify the various types of law to find the various types of social solidarity corresponding to them. Hence it is probable that there is a type which is a symbol of that particular solidarity of which the division of labour is the cause. Therefore all we have to do to estimate the part taken by the division of labour, is to compare the number of juridical rules which express it with the overall extent of the law" (*ibid*. p. 23).

As we see, to get objectivity, we have to substitute in place of the idea we form of things in the abstract, the reality that experience and history oblige us to recognize as being due to them. So in this way only will sociology escape building on abstractions and will scrupulously heed all the ties with what is real, as revealed to it by studying the nature of morals. Such are the links, as seen in these lectures, between professional ethics and economic evolution, between civic morals and the structure of the State, and between the morals of contractual relations and the juridico-social structure in all its variability. Such also—as in the lectures still unpublished—are the ties which link family sentiments and obligations to the variable forms of the family, and which link these in turn to the diverse structures of societies. In short, we have solidarity, the value attached to the person, the State, classes, property, contract,

1

exchange, corporation, family, responsibility, etc.—all these are given phenomena, no matter whether material or immaterial; they are presented to us each with its own nature, which we have only to take as it is, in its shifting complexity, so often clothed with a deceptive simplicity. In face of these phenomena, let us keep clear not only of any arbitrary interpretations, but of a too facile and tempting approximation that seems to account for the phenomena from the start either *a priori* or by those supposed constants in human nature, instinct and need. Reference to a nature which seems to protect us from the arbitrary is not enough to give us a true objectivity. Whilst nature gives form, history transforms. Observation serves us only from an angle of what is relative and when it places the fact observed within its own conditions of existence. That setting, like nature indeed, includes compatibilities and incompatibilities, upon which the balance and play of the function depend. But this balance itself is but a stage in the act of becoming, and the adaptation of the function is not obtained from the start or amenable to the interpretation on horizontal lines alone through the environment or ambience of that particular moment. The function is given preparation by sequences in time on vertical lines. The given social reality, which we should not construct but observe as a thing, must therefore be observed in experience as well as in history. It is the working of the function alone which may be observed in the actual present.

But the working of the function is not the function, any more than the function is nature. These three elements are distinct and all three are to be observed as assumed to be within time, and, let us repeat without any ambiguity, are to be "treated *as* things".

This, then, is required by the theme of science, that we described as demanding the method of sociology. But the theme of 'the social' which posits its existence also has its demands. It remains to be seen if and how these may be reconciled with those of science.

The claims of science that forbid any crossing of the bounds of immanence, by that very fact favour the concept of the 'normal', as distinct from that of the 'pathological', and this concept of the normal by force of this difference can serve as a criterion to weigh the reality observed. One even sees this notion of the normal fact or type being substituted for the notion of the ideal or what-ought-to-be, and being presented as suitable to govern our conduct instead of being confined to throwing light on the means. In this perspective, a phenomenon would be presented as normal if, to begin with, it appeared as sufficiently general in any given society where it formed an average type ; above all, and in a deeper sense, if it presented an exact correlation with the structure of the society from which it issued. This correlation which provides the basis of the normal is more than the generality—itself hardly other than an indication. On this definition, normality constitutes the soundness, identified with the well-being, of the society and thus is destined to guide its efforts towards adaptation. On this point we cannot help remarking that the generality may be a deceptive indication: that is, if it be possible that a conduct for survival (i.e. one that remains the same despite a modification of the structure, to which the conduct reacted normally) can preserve its generality for a certain length of time. Likewise we may observe that the precise conformity of any conduct with the correlative structure is a thing very difficult to appraise: this emerges from the examples cited by Durkheim in his chapter on the difference between the normal and the pathological, some of which seem pretty arbituary. This difficulty is aggravated by the fact that each normal type is normal only to one definite society and not to human society in general. It is increased too by the fact that, to establish the existence of the normal type, there must be a classification of societies. There is a scheme of this classification in the *Règles*, but it errs rather by over—systematizing; in its mechanical and *a priori* character it runs counter to the relative point of view which, by the principle of correlation, it should have adopted, and should have allowed to apply.

Let us assume that the structure taken as a criterion of normality is, as it should be, indeed that of a given society, clearly defined, located and dated. Who will then assert that the system of beliefs and behaviours, the mentality and the institutions, which ought normally to arise within it and make their force felt, are necessarily determined on that premise? When the structure is invoked as a criterion, is there only one possible answer? Why should adaptation—for this lies at the root of the question—why should this not comprise various modalities?—It could be guided perhaps, at least in part, by the more or less conscious desires or choice of the human agents, nevertheless, who collectively or individually make the adaptation work. Likewise, it may happen that a geographical site may in one instance impose the town on the man, and in another, that man may impose the town on the site.

Reference to the clearly defined normality alone, as we said, keeps us in any case, in company with Durkheim, within the strict limits of experience, and excludes any appeal to transcendency. The causal link by which it is sought to establish mechanically, as it were, a correlation with every social structure, derives therefore from the imperative theme of science, that we have already defined; and thus seems to lead sociology, which has evolved under this sign or indication, to a pure positive science. Nothing of the sort, however. Durkheim was not long in revising his earlier view when he approximated the distinction between the ideal and the real to the distinction between the normal and the pathological; this view, from the time of his most rigid orthodoxy, went hand in hand with the assertion, set out above, which limits the positive science of that view to a remarkable degree: that is, the assertion of the specific nature of 'the social', having regard to the psychological and, more especially, to the biological. Could we not simply say that the type here propounded, unlike that explained by mechanism or positive science, excludes a reduction to simple elements as well as the claim to proceed always from the lower to explain the higher. Sociology, the subject of which lies in nature and not outside it, ought to be a science as natural science is. There is, however, a difference: it has to keep being to a science and yet let nothing escape it of the quality peculiar to 'the social' —its own subject and concern and one that at the same time can be no other than a human subject. That is because the social phenomena which sociology apprehends are the phenomena of human societies, and because, according to Durkheim, it is through his social character that man becomes progressively more human. And this is convincing because sociology is able at will to proceed from man to rediscover in analysing his nature the presence there of the society, or to proceed from the

Introduction

society, the study of which will inevitably take him on the road to Man. 'Man-in-society', or 'society-in-man',: the two formulas are equivalent and may both serve to define sociology, if it be true that man has of necessity a social dimension and society—no less of necessity—a human composition.

In this way it is possible to temper the strict positive science of this distinction between the normal and the pathological, which was given a kind of exclusive warrant by the guiding principle of science, both for defining objective knowledge and for providing action with its ends no less than its means. And the bondage to positive science becomes less onerous still in the degree to which the ideal is, in consequence, the more sharply distinguished by Durkheim from the purely normal. Then, the collective consciousness, considered more and more in its nature and in its action as a consciousness, would cut adrift from the morphological structures with which it began by being closely bound up: it would gain height and take on the character of near-universality, and in the end assume the function of transcendency in its role (increasingly clear), of being a focus of the ideal.

There is, then, no rigorously strict method that will serve: the human is not to be absorbed in any kind of mechanism or materialism. But the human, by virtue of its social dimension, and to the gain of consciousness, is only saved at the expense of the individual. At this point we see the methodological tyranny of the theme of science re-appear. Under pressure from that tyranny, we get the exclusiveness given to the interpretation by causes entirely social, an exclusiveness to be equated with that given above to the notion of the 'normal'.

It is in fact the peculiar character and, one must say, the narrowness too, of the Durkheim sociology that—once the social dimension of man is recognized—it seeks to keep that dimension alone to define humanity, and for the alleged reason that the social dimension alone can be objectively apprehended. Hence it follows that the specific nature of 'the social' that we laid down as a major theme alongside that of science, in one sense limits the privilege of 'the social'; in another sense it gives it fresh force, since it provides 'the social' with a veto in respect of all that is pure individual spontaneity, the essentially subjective nature of which rules out any objective determination. The author thus seems to believe he must sacrifice the individual to 'the social' to allow 'the social' to save the human in the face of science.

It is a sacrifice however that, like Abraham's, is not made without stress, hesitation and compromise. We can judge of this by the place and the role assigned to individuality: here we see, alongside the will to restrict, not to say deny—a will of course very often and clearly asserted—a tendency here and there and by degrees to become less prohibiting. Hence, whilst we find a clear inclination to close up the Durkheim theory within its own exclusive and rigid sociality, there is an equally definite trend to open it, rather reluctantly perhaps, but more however with the motive of carrying it further than countering it. Let us look at this more closely.

First, we have no wish to deny the censorious conclusions that abound in the *Règles de la Méthode Sociologique;* that is natural enough in a chart of method, objective and therefore severely scientific. Anyone capable of declaring that 'every time that a social phenomenon is directly explained by a psychological phenomenon, we can feel certain that the explanation is wrong", is led naturally (even if he grants that one cannot make an abstraction of man and his faculties), to maintain at least, and to emphasize that "the individual could only be that indeterminate substance which the social factor determines and transforms". The same kind of logic leads to the assertion in respect of sentiments, that "far from being the foundation of social organization, they are the result of it." This does not go as far as sociability, which our rigid sociologist, from the same point of view, would deny to be due, initially, to the instinct of the individual; in its effects he would place it to the credit of the influence of social life. Human individuality cannot be said to be obdurately set against life in society but certainly, no less, it does not take to it willingly: it is, then, only an indeterminate and malleable substance which would be no more capable than the Aristotelian substance of proceeding to any action of its own accord. This is so, since its passivity appears to be devoid of any true principle of determination. Here, too, we have correlatively, as in Aristotle, a science only of the general, that is, of the social type, or, as described above, a science only of the social dimension of the individual. There must indeed be nothing ambiguous about the meaning of the word 'general'. For if we take 'general' in the sense no longer of 'generic', (as we have just done, by analogy with Aristotle), but in the sense of 'indeterminate', the 'general', taken thus, can be used on the contrary—because it is synonymous with 'indeterminate'—to qualify and banish individuality. The *Règles* seem in fact to lay down that, although the general qualities of human nature engage "in the work of elaboration that results in social life", their contribution "consists exclusively in states of being (*états*) of a very general kind and in tendencies that are vague and hence malleable, which in themselves could not take on the definite and complex forms that distinguish social phenomena, were it not that other agents intervene."

It should be understood that these other agents are the social factors, since, as we are told once again, to explain this by way of the individual would only allow the specific nature of 'the social' to elude us. Let us then see what, finally, we are asked to think of the so-called individual psychological propensities, that are constantly invoked to explain everything: this is, that "far from being inherent in human nature, they are either entirely lacking in certain conditions of society, or these conditions present so many variations from one society to another, that the residuum obtained (after eliminating all these differences), which alone can be looked upon as of psychological origin, is reduced to something vague and schematic: this leaves the facts that have to be expounded infinitely far off from us."

According to Durkheim, however, a more immediate and precise effect can be claimed for psychological phenomena of an individual kind, when capable of bringing about social consequences: this is when they are so closely linked with social phenomena that the action of the one is inevitably merged with that of the other. This applies in the case of the official whose standing and efficiency lie in the social force which he personifies; it also applies to the statesman or the man of genius, to whom he concedes only that 'they derive from the collective sentiments—of which they are the focus—an authority that is also a social force and which they can place, to some extent, at the service of personal ideas." And as if this tiny concession were still too much for him, Durkheim hastens to add, rather disconcertingly: "But these instances are due to individual accidents and hence they are unable to affect the constituent features of the social type which alone is the subject of scientific enquiry." (*Règles*, 2nd Ed., p. 111, note I). Finally, to forestall any false hope he might raise amongst the individualists, he confirms this not very encouraging remark by a conclusion which is even less so: 'The reservation to the principle laid down above is thus not of any great importance to the sociologist." And so, by the veto of the same old censure by methodology, stubbornly set against all possible subjective backfire, we see how any inclination to temper the rigid exclusiveness attaching to the purely social interpretation is inhibited.

The individual cannot break the thread of this interpretation to cut in with his own causality, even by way of contribution. His reasoning, it is true, cannot be dismissed. But this reasoning could only contribute its enlightened adherence and never a creative effectiveness to the interpretative scheme or outline devised with the help of social factors and the needs of structure. Such is the limited role of our autonomy, which is defined on the same narrowly rigid lines in the *Education Morale* as the process of a lucid and deliberate recording but not of law-making: "there can be no question of regarding human reason as making laws for the physical world. It is not from us that it got its laws . . . it is not we who made the design of nature: we re-discover it through science; we re-think it and we understand why it is what it is. Therefore we can submit ourselves to it as far as we are convinced it is what it ought to be, that is, as implied in the nature of things; and this we can do, not simply because we are constrained to by material forces, but because we judge it to be well to do so." And from this analogy with the free (because rational) Stoic adherence to the cosmic order, Durkheim concludes: 'In the moral order, there is room for the same autonomy and there is room for no other." (*Educ. Morale*, pp. 130–2).

But we have to go to the very end of Durkheim's analysis of the autonomy of reason as he defines it, to understand that he assigns no specific role to the individual as such. This he often asserts, and we can trace it in the well-known statement on the determination of the moral fact. (*Sociologie et Philosophie*, p.

95 *seq.*): "In the sphere of morality as in the other spheres of nature," we read, "individual reason has no particular prestige as such. The only reason for which one can legitimately claim —there as elsewhere—the right to intervene and to rise above historic moral reality with the aim of reforming it, is not my reason, nor yours, but impersonal reason, which is only truly realized in science. . . . This intervention of science has as its end the replacing of the collective ideal of to-day not by an individual ideal, but by an ideal which is also collective and which expresses not a particular individuality but a better understood collectivity."

As we see, the position taken up could not be clearer or more categorical. And so that there should be no mistake, the author, going to the extreme limits, adds: "Could it be said that society really attains this highest consciousness of itself only in and by an individual intellect? Not at all, for society only reaches this highest consciousness of itself through science; and science is not a thing of the individual; it is a social thing and essentially impersonal." And finally:—"If we are to understand that reason in itself possesses inherently a moral ideal which is indeed the true moral ideal and which reason could and ought to set against that ideal pursued by society throughout history, I say that this is *a priori* reasoning, an arbitrary assertion in contradiction with all the facts."

This would indeed condemn, clearly and unconditionally, any truly individual initiative. It is, too, a limitation, far more Stoic than Kantian, of the autonomy of reason—once its part is recognized and admitted. Is this condemning and limiting however, for all their preciseness, free of ambiguity, and serious ambiguity at that? We cannot think so. To begin with, there is this approximation (certainly more apparent than real), between what is termed here "impersonal human reason, which is truly realized only in science," and science itself, " a thing not of the individual but a social thing and essentially impersonal." This approximation is evidently intended to convey to us that science, and reason that goes to produce it, are impersonal and in consequence objective, only because they are social things.

Had not Essertier, in his book on *Formes inférieures de l'Explication*, already condemned precisely this confusion of the collective and the impersonal in respect of both science and reason? "Impersonal thought," he writes, "is by turns thought which is that of no individual in particular, and objective thought, on which the whole system may hang: impersonal thought is true thought but it is also collective thought. Thus, collective thought has indeed created true thought. In point of fact, what is expressed in the impersonality of true thought is the whole of the personality. It represents no less than the victory of the individual over his own subjectivity. Now this subjectivity is made up precisely and above all of collective representations. In brief, the impersonality implicit in the truth assumes in the one who discovers it or sets it forth in full knowledge of the case, a very high development of the personality and complete emancipation from collective ways of thinking . . . in order to give place to the object, that is, to the impersonal."

This criticism is quite right. For if impersonality, which is in fact both indication and criterion of objectivity, has indeed its part to play in science, this part is not to discover the meaning but to approve its accuracy through collective support—given to it or not—by the scientific community. The discovery of the meaning, on the other hand, rests very much with this or that scientist. Such discoveries, however, are not likely to be simultaneous. We should not therefore, on the fallacy of a collective approval of objectivity, dismiss the scientist for the benefit of society. Moreover, if science is collective in effect and impersonal, this is due more to the sum of individual discoveries submitted to a general verification rather than to collective acceptance that serves as formal approval of the individual discovery. Science, it is true, is not the thing of one individual, as Durkheim rightly observes. But the fact that it is the thing of several individuals does not imply that it is non-individual, and therefore a social thing as opposed to the individual. Thus we reach the final argument. To assert the role of a purely individual reason in the interpretation would amount to making a kind of monad of this reason. This monad would from the start contain in itself the whole of the interpretation or of the ideal propounded. The reproach of being arbitrary is then too easily turned back upon the argument that makes it. Bachelard has demonstrated convincingly that science is not all reason, and that rationalist activity is not fruit-ful except when it is at grips with experience and in constant debate with it. This would be a debate started by its individual initiator, who anticipates rather than opposes the collectivity of those who will go over it again to verify or correct it. Why, precisely where there is need to verify, should we demur to subjectivity in individual reason, in favour of collective reason —said to be alone scientific? As if this very collective reason were entirely safeguarded from the subjective perversions, and as if individual reason (most often the very reason that is creative or inventive), were on the other hand inevitably sus-pect of arbitrary subjectivity.

The fact remains that one inflexibility leads to another. Thus, the first stringency of methodology, reinforced by our own first theme of science, seemed to entail, along with the condemning of the theories of final cause and animism or psychism, an exclusive criterion of normality. Further, it seemed, a second stringency, deriving from the first, gave an exclusiveness to the interpretation resting on social causes alone, whilst any indi-vidual cause was disallowed. And now we see a third stringency, deriving in turn from the second, coming along to throw dis-credit—apparently, at least—on the historical interpretation, a discredit similar to that thrown on the individual interpreta-tion. There is a lapse here, all the more curious since it is not irretrievable, and since, on the other hand, it shows the danger of an excess of logic. This is shown by an analysis of that internal *milieu* which remains the only territory where the interpreta-tion by social facts can be sought, once the individual factor is discarded, as we have just seen happen.

Defined thus, indeed, this internal *milieu* would exclude any individual factor and would only include the morphological elements of structure, depending on the way the constituent parts of society are attached to the soil or grouped among themselves. In short, what is here at issue to interpret the social processes is, for the society in question, its *present* conditions of existence and the motive powers active within it *at the present time:* this means, according to Durkheim, the demographic volume and density, to which must be added, for any given time, the influence of any neighbouring societies. These causalities, by the way, may be observed at more than one level, for there is not one social *milieu* to consider but all those, of the family or of others, which exist within the society in question.

"This concept of the social *milieu* as a determining factor in collective evolution is," Durkheim declares, "of the highest importance: for if we reject it, it is impossible for sociology to establish any relation of causality. In fact, if this type of cause is set aside, there are no concomitant conditions on which the social phenomena can depend." We see how the emphasis thus laid on the concomitance (with which the causality alone is linked), excludes any sequence and hence any historical interpretation; and we see, too, how the exclusiveness of what Durkheim calls the *circumfusa* leads, in the *Règles*, at any rate, to the disqualification of the *præterita*. But why this censure, which the author, in his own research, would be the first to reject? We have spoken of an excess of logic that fetters one stringency to another. To that should be added a kind of terror of the philosophy of history which (since he had seen this philosophy as the undoing of Auguste Comte) leads him to banish history itself at the same time. One would have to be unfamiliar with the whole work of Durkheim to take literally this statement—which expresses what we have just said: (*Règles*, 2nd Ed., pp. 116–117) "One can well understand that the progress achieved in a given period . . . makes new fields of progress possible. But how does it pre-determine them? . . . One would have to admit an inherent tendency that simple humanity ceaselessly to go beyond its achievements . . . and the object of sociology should be to discover the manner in which this tendency developed. But without harking back to the difficulties such a hypothesis implies, there could not be any-thing causal about the law which expresses this development." Let us admit that this alludes to Spencer's pseudo-law of evolu-tion or to Comte's law of the three states (*états*). But after having cut out (in subsequent lines) 'the motivating faculty that we imagine as underlying movement', how comes Durk-heim to make the error of asserting, and without reservation, the following principle: " The antecedent state does not bring about the consequent state, but the relation between the two is exclusively chronological." What should we think, then, of the economic crisis of 1929—to take one of many instances? Must we believe that if we open our eyes to the concomitant conditions, we are obliged to shut them to antecedent conditions? As if the necessary horizontal solidarity of the conditions of existence of the present time ought to detach itself from the vertical solidarity that links them to the preceding equilibrium; as if function were to owe nothing to origin?

Taken literally, the whole passage we are considering would have us believe that we could not seek for causes in history without admitting them to be tied by the stringency alone of a single law which determined them all. We read: "If the principal causes of social events were all—(but who would say 'all'!) —in the past, each society would be no more than a perpetuation of the preceding one, and the various societies would lose their individuality to become no more than the different moments in time of one and the same development." Does not the individuality in question, on the contrary, depend precisely on the historical variability of the conditions of existence? History may indeed not be all, but that is not to say that it is nothing.

Moreover it is to Durkheim himself that we look to put Durkheim right. What in fact do we read in the first few lines of the first lecture now published? This: "The science of morals and rights should be based on the study of moral and juridical facts. These facts consist in rules of conduct that have received sanction. The problems to be solved in this field of study are : (1) how these rules were established in the course of time, that is, what were the causes that gave rise to them and the useful ends they fulfil. (2) the way in which they operate in society, that is, how they are applied by individuals. . . . These two questions are distinct, but even so . . . they are closely connected. There are the causes which have led to the establishment of rule, or law and order, and there are the causes responsible for the ascendancy of rule over the minds of men, sometimes of few, sometimes of many. These causes are not identical but are yet of a kind to act as a check on each other and also to throw a light one on the other."

What has happened that it should seem possible to look to Durkheim himself for the means of refuting Durkheim or at any rate of correcting him? No doubt he was a victim of the relentless stringency of a reasoning that was more concerned with excluding the doctrines he rejected than with following the reality he wished to expound. Hence we have what we termed above a chain of stringencies, each one producing the next. We should add that this chain is matched by a series of approximations and antitheses that very quickly become controversial: such as the antithesis between the objective and the subjective, which is but another form of the antithesis between mechanism and final cause or again between the scientific and the mystic. Hence the exclusion of the individual, as if he must always be identified with the purely subjective, and not amenable to any determination. This leads naturally to the antithesis of an individual taken in this sense and excluded from a social *milieu* which alone has privileges. There follows an approximation of the *milieu* with the ambience, then one of the ambience with the concomitance and of the concomitance with the present; this leads finally, by antithesis to this present, to a dismissal (almost by pretermission) of the past. This past is seen in the form not of ordinary sequence, so naturally complementary to the concomitance, but in the form of a whole, ostensibly directed by one single law. And thus we see how, in the final reckoning, the recourse to history for help with the interpretation is condemned—and for reasons that are valid only for the philosophy of history. It is condemned, we might say, with an impatience that cuts across any logic. All the same, he quite readily goes back on himself. Indeed, does he not add this statement, very different in spirit: "The determining cause of a social fact must be looked for among the *antecedent* social facts (our italics), and not among the states of the individual consciousness." (*Règles*, 2nd Ed. p. 109).

With the reservation of this principle that restores the necessary balance between interpretation by *milieu* and function, and interpretation by history, we may as well admit this precept of a proper sociological method (with a hint of finality, all the same, that Durkheim would reject but that seems essential): "The fitness or otherwise of institutions for their purpose can be established only by relating them to a given *milieu*," and since these *milieux* vary, "there is a diversity of . . . types which, qualitatively distinct from one another, are all equally grounded in the nature of the social *milieux*." (*Règles*, 2nd Ed., pp. 118–19.)

We should not blink the fact that Durkheim has been reproached with having broken away early from a methodical analysis of types varying in *milieu* and morals; and hence from the setting up of a suitably experimental typology of groups, and also from any correlating of their institutions to their particular structures—all in order to become to some extent the metaphysician of sociology. He is said to have gone over—an unmistakable sign of it—from societies in the plural and collective representations to a Society in the singular and the collective consciousness. Certainly there is no doubt (as I have written elsewhere and shall revert to further on), but that the collective consciousness has gradually been acquiring a halo and becoming, as it were, personified; it has, we might say, been climbing up to assume ever more clearly the role of a real focus of the ideal. There is also no doubt that Durkheim saw more in sociology than a science of societies, and that he thought that sociology must find its fulfilment in philosophy, a philosophy with a positive basis, however. He may have hoped too much from the collective consciousness and have magnified and even deified society. These ambitions, and perhaps these illusions, related him to the doctrinaire side of Comte which too often obscured the other and made him fail to recognize it. They never made him underrate, however, or abandon the meticulous study of societies and institutions in all their variety and the different modalities of their existence, as well as of their functions and the working of those functions. He never ceased to declare that the system of Morals he wished to establish on a solid basis called for much preliminary research on the various ‚roups and on the life and part played by those sub-divisions of all kinds that may exist within them.

These lectures on the nature of morals quite rightly start off with this diversity provided by nature and developed by history, in trying to determine the forms of moral conduct in terms of the manifold types of societies and institutions to which they correspond.

It might be said that the course on professional ethics is just as interesting from the point of view of method as of doctrine.

As to method, we find in the lectures an analysis of the function of professional grouping; it does not exclude, but rather, calls for the study of the *milieu* considered against its present conditions of existence (*circumfusa*) and parallel with that, of the origins sought from history (*præterita*) and from ethnography. We also see statistics that reveal the working of the function. And then we are perhaps especially aware of the author's concern to draw lessons from experience and history for organizing the present. So we find professional ethics linked with the very nature of the social grouping by the guild or corporative association—from which they govern the behaviour of individuals.

To Durkheim the problem is therefore one of finding out what the corporations should be if they are to be in harmony with present conditions of existence. "It is clear," the author says, "that there can be no question of restoring them in the form they had in the past. They died out because they could no longer survive as they were. But then, what is the form they are destined to take? The problem is not an easy one. To solve it, we shall have to be a bit methodical and objective, so we must first arrive at how the corporative or guild system evolved in the past, and what the conditions were that set the evolution going. We could then judge with some assurance what the system should become in future, given the conditions at present obtaining in our societies."

We therefore have to distinguish in the institutions—corporations or other—between constants and variables. The first correspond to the permanent role of those institutions which seem to be constituent parts of any social framework; the second correspond to forms of adaptation brought about by changes in time or *milieu*. To give a true estimate, we may recall that all that Durkheim has said on the subject was written long before the various present-day experiments with corporative associations or trade unions and new versions of the old system. Although these experiments may have been made in a wrong-headed attempt to monopolise the corporative system and so to subordinate it and alter its form, they still give no proof at all that the particular role of that system can ever disappear. The existence of more closely integrated and specialized groups must always seem normal and necessary to govern interests and lay down rules professional and moral, which the central authority could only direct from too remote a point and without the adequate skill. Is it not significant here, that public law (*droit public*) is seen crumbling away in so many forms and delegating part of its authority, even though it must keep its sovereign right of arbitrament?

We have just passed imperceptibly from the standpoint of method to that of doctrine, and this is roughly reflected in the lectures themselves. Here, the interest lies with the economic side and for two reasons: we are aware at all points of the out-line of this division of labour in which the author sees, far more than an economic phenomenon, one that is intrinsically social, the inevitable result of social causes (variations in volume and density of the grouping). We reach the conclusion that the continued progress of the division of labour has not carried with it the normal correlative of integration or regulation. The professional grouping or association thus appears as an essential active principle of solidarity: it has, however, not quite acquired the complete identity necessary for playing its proper role.

And it is here we come again on one of the main philosophic themes of Durkheim: the need for the individual—who is what he is only in the context of society—not to be deprived of any of the framework or supports that the various social groups or sub-groups have to offer him, if he wants to escape disastrous anarchy. In rejoining this main theme we come also to the governing idea of his book on suicide and of the lectures on respect of the person. Organize, organize—and in so doing, you will raise the moral standard. In this, Durkheim finds his way again to one whom he often severely criticises, that is, Auguste Comte, with whom we linked him to begin with. But for Durk-heim, the development of professional associations entailed direct political applications both in the national and inter-national spheres. These prompted some remarkable prophetic forecasts on his part and also, it is true, certain illusions. But we must say again that these lectures would not be framed in the same way to-day as half a century ago: so we may be justified in letting the forecasts stand, but it would be unfair to reproach him for the illusions.

The Second Edition of the *Division du Travail Social*, with its striking introduction is as we know given up to the professional group forms, but a far greater interest attaches to the lectures in this volume. I am not claiming to give a summary of the lectures and thus make the study of them unnecessary. I will only point out the importance of those (the greater number) which treat of civic morals and which, in order to lay down these morals, analyse the nature of the political society and of the State. In these, we touch on that part of the book which is most fertile in ideas and the most unexpected, too. It is not surprising to find the State first and foremost and as if provisionally, linked with the idea of constituted authority and the juridical set-up of the group. But as the State is the chief organ of the political society, we must begin by defining the society. Durkheim, we assume, would not define it by starting with the domestic society, the family, in which he declines to see its origin. Nor would he see it in the fixed attachment to the soil— for there are nomad societies—or even in the numerical size of the population, although this is a factor. No, he defines it by the fact that the political society comprises within itself secondary groups of different kinds, which are of service to it rather than disservice—if it is a fact that they are indeed its constituent parts, and that the society itself can never in turn become a secondary group. It is within a system of federation alone that we can see the basic and primary aspect of political societies, concurrently with a secondary aspect reflecting that part of them which is federated—that is, the part devoid of sovereignty. Short of this, if I am right, a society is termed political when it appears (to avoid Durkheim language) in the form of a sovereign all-embracing entity. As a counter to this definition (that clearly reflects Durkheim's thought whilst shewing its inconsistency) someone, we expect, will bring up the well-known simple segmental society, which is seen in the dangerous chapter of the *Règles* already referred to as the source of any political set-up and the basis of classification of the various kinds of society. But it is right to recall that the matter is presented above all as a hypothesis; this, it seems, tempts us to see in this simple political society so-called, a kind of limit in the sense (if you like) that Bergson advances his 'pure perception' as a limit never reached, where consciousness and materiality meet. Here, too, if we are to define the nature of the political society, we should start with 'the all-inclusive' which would be, as it were, the differential of political synoecism. To this comparison, however, there has to be one reservation : the emergence of the 'all-included' non-political parts and of the political 'all-inclusive' is not in sequence but simultaneous.

The State, then, is the organ of a complex group of this kind but is itself provided with secondary, executive organs; this does not mean, as we might think, that we must look on the State first and foremost and mainly, as executive. Certainly it is not that; it is not even an organ in the strictly juridical sense. Rather, it is a representative, collective *'brains-trust'*, as we might say nowadays, whose proper function, coupled with autonomy, consists, in Durkheim's phrase, "of elaborating certain representations that are valid for the collectivity", and, of course, also of directing its common interests in the name and in place of the collectivity. The State would therefore be directly deliberative, and only indirectly executive, and executive by proxy given to the government. Does there in fact come about a kind of universal planning of thought and behaviour, in which the few, making up the small collectivity *sui generis*—for such it is —having the name of State, would do the thinking and willing for all? The answer is 'No': that is, if, as Durkheim thinks, it is true that the inherent rights of the individual (those we might be inclined to set up against the miniature Leviathan) are not inherent but, on the contrary, bestowed by the State on the individual: and bestowed to the precise degree that the separate outline of the individual becomes more clearly etched on the social background, in the natural progress of social life which advances from heteronomy to autonomy. The individual then draws from his habit of obedience the aptitude to command and to get recognition as a mandatary of those governing and as a moulder of the society which first moulded him.

All the same, we must not believe that Durkheim would follow us up to the point of individual autonomy to which we should like to carry him. We read in these lectures the very clear statement that was quite enough to hold him back: "At the same time that the society nourishes and enriches the individual nature, it tends inevitably to subject that nature to itself. Precisely because the group is a moral force and to this extent superior to that of its constituent parts, the group tends inevitably to subordinate the parts to itself." Let us hasten to add that, happily, we are told almost in the same breath that as a society expands, its pressure relaxes. At such a juncture the political society becomes tutelary, because its pressure bears less direct on the individuals than on the secondary groups: it has on the one hand to keep these groups in a state of balance, and on the other, it has in its own interest and as a duty to defend the individuals against the groups. If the individuals, threatened by the more immediate and therefore tighter grip of the small groups, were held in servitude to them, these groups for their part, strengthened by the bondage they impose, might well turn dangerously against the political authority of the State that hems in their restive feudalism. As the vassals would then become the masters, the political society itself would go by the board. It is from that society's instinct for self-preservation that we get the mobility and freedom of the individual. There is, as it were, in all this some mechanism of counter-poise that recalls the Montesquieu to whom Durkheim dedicated his Latin thesis.

This means that we are not dealing with a *mystique* of the State: and if the State, as Durkheim holds, is led to a continuous expansion of its functions, it is because social life could not become more complex or more varied without increasing its control by rule. This structure of rule would protect the individual far more than fetter him—and for the very reasons we have just examined concerning the relation of the State to its sub-groups. Durkheim returns to this several times. What lies at the basis of the right of the individual is not the notion of the individual as such, but the way in which the society conceives the right and the valuation it puts on it. He goes on putting this estimate always higher, but he is so far from seeing a threat to the individual in the State that he assigns to it, on the contrary, the role of "calling the individual by degrees to a moral way of life", and that, at a time when religious and moral beliefs seem to him to be declining, he lays stress on the cult of the human person which, meanwhile, is in the ascendant.

This rise of the person, who, emerging from the indivisible nature and common ownership of primitive communities to get himself by degrees recognized and respected, is at the same time the dominant idea on which the lectures devoted to property and the contract turn. It is an easy matter to give a summary idea of this. First of all, how does property come about? The sacredness diffused through all things, which originally shielded them from any profane appropriation, was brought by means of the prescribed ritual to the threshold of the house or to the confines of a field and formed, as it were, a zone of sanctity which protected the area thus bounded against any alien trespass. Only those could enter that zone who were qualified by ritual bonds to have contact with the unseen powers of the soil. Then, as the sacredness passed from the things taboo to persons mystically qualified, it was these persons, in turn, who were able with exclusive right to bestow on the things they declared to be *theirs* that sacredness which made them a protected property. And so we get a passing from collective property into individual property, which is truly established only from the time when the individual becomes able to impose the prestige of his person by presenting it as the very embodiment of his ancestors as a group and of his sacredness or potency. In those days, the pattern for private law (*droit privé*) was sought in public law (*droit public*), whereas to-day public law could find many models in private law, and the behaviour of individuals might often provide a model for the conduct of nations.

Turning to the contract, research has disclosed comparable filiations or forms of derivation. These make us understand what metamorphoses and what extraordinary and costly complications have gone to getting a result so apparently simple as the free reciprocal undertaking of two individuals by simple declaration and acceptance of their declared will. Neither the intention nor the pronounced declaration serve for this in the beginning: the ritual has to be set in motion, the whole ground of the *right* dug over, as it were. And it is not even single persons but whole groups that, collectively, are confronted with a process as complex and costly as a military operation or some hammered-out peace treaty. For single individuals to be brought in—according to their own particular needs—and to ensure that the object or pledged word exchanged should make them from the outset owners or sworn parties, there still had to be, at one and the same time, a whole series of simplifications, as slowly acquired as are those of habit inherent in instinct. More than that, to take advantage of these simplifications, there had to be that advent of the person *qua* individual, an idea as a force, that we have already come across, and by virtue of which in fact this person would be empowered in the contract to pledge his word voluntarily and effectively.

What is the final conclusion to be drawn from this summary review? In no matter what region we have traced the paths of Durkheim's doctrine, we have seen them converge towards exalting the human person until it becomes in the end no less than a cult. On the other hand, in analysing the method and the main themes that guided him, we seem to have established that this method—carefully framed to let nothing of the specific nature of 'the social' escape, any more than of 'the human', (social and human being, in his view, one)—was steadily applied to avoid any intrusion of the individual, either as an initiating agent or as a factor in interpretation. The danger of the individual coming into the system seems to be that with him an incalculable and capricious subjectivity is introduced, and obscure tendencies that are unconscious or mystic—not to speak of pure fancy—which sever the train of reasoning and defy explanation. All these, it might seem, would be fatal to the objectivity that the new science has to achieve at all cost but without losing sight of its object.

How are we to escape this apparent paradox? How avoid the dilemma without disintegrating this human creature so conveniently determinable in his social dimension, while still casting out the individual offered up, inevitably, as it seems, to science? If we are satisfied to keep the social dimension alone, as proposed in the method, would this be enough to allow of the doctrine—as we have seen above—exalting the person? Yes, but on the sole and indispensable condition that this person be no more than a reflexion of society—its first-born, so to speak—and that notwithstanding all his conformity, he shall become so far a man of parts or talents to allow him to shine with his own light and to win respect as a person. That being so, this tutelary society, as we already indicated, has had to be lifted far above the average level of the psychic nature, and of the morality and creative genius of individuals in the mass.

In his book *Le Suicide* (pp. 359–360) we come to the following passage: "It is a profound mistake to confuse the collective type of society, as is so often done, with the average type of the individuals who compose it. The morality of the average man is only of very mediocre strength. It is only the most essential principles of ethics that have made any great mark on him; even these are far from being as precise and authoritative as in the collective type, that is, in society taken as a whole. This confusion, which is the very mistake committed by Quételet, makes the origins of morality an unsolvable problem. For since the individual, as a general rule, is so mediocre, how have morals come to be established which so far surpass the individual, if they express only the average of individual temperaments? The greater cannot come out of the less." In face of this inadequate average, morals are presented to us "as a system of collective states" (*états*).

In the *Education Morale* we find the same note but with even greater emphasis in the idealistic sense (p. 140): "The society that we have made into the objective of moral conduct goes incalculably beyond the level of individual interests. What we should above all cherish in it is not its body but its mind. And what is called the mind of a society is no other than a collection of ideas which the individual in isolation would never have been able to conceive and which outstrip the limits of his mentality, and are shaped and given life only by the coming together of a great number of associated individuals." In one work after another, we see society winning fresh claims to noble qualities. Here it is as presented in the well-known passage on '*Judgments of Values*': "The fact is that society is at one and the same time a legislator to whom we owe respect, and the creator and trustee of all those benefits of civilization to which we cling with all the forces of our soul." And last, the main point and climax of the same text : "It is society that thrusts forward the individual or obliges him to rise above himself. . . . It cannot be constituted without creating ideals."

If we step down from this climax to the point at which we set out, we can but note that in this comment—hard to dispute— there lies the following: the very fact of the aggregation of individuals in society (with all the systems of structure and equally, all the mental interactions and mutual behaviours that it inevitably involves), gives rise to a whole network of representations, symbols, exchanges and obligations unknown to the individual in isolation. How, we may ask, could a society so richly endowed have taken any lesser form than that of a collective consciousness, in which Durkheim (always with the same scientific mistrust of any individual subjectivity) finds it natural to place the source of ideals and the basis of all regulative order? Does not the collective nature of this consciousness keep it within the compass of the objective, and does not its synthetic character ensure for the consciousness, along with the necessary specific quality, the creative power it seeks? If we agree, it means that to socialize is to humanize, without any offence to science.

But what if the whole of the human did indeed not reside in the collective consciousness? Suppose that some part of the human could on the contrary only emerge on an individual lodgment, having in mind of course the individual outlined against the social background, the sole place perhaps in fact where he is able to come to life, but where nothing prevents his acquiring in turn the capacity for synthesis and inventiveness, and the force of significance and obligation. Perhaps in that case such an individual, from the aspect of the person thus engendered in him, would no longer be satisfied with having a delegated character and with having to look at an image not his own. With the reflexion giving place to his own light and once the strands of a true individual consciousness had been knotted in him—also a synthesis—perhaps this individual would consider that a consciousness such as his should be called —not the sole source, but one of the two sources of becoming human. Perhaps, too, he would claim inclusion by the same right as the other source, in the system of interpretation in terms of sociology; that is, if it be true that, in face of the individual thus presented to us, we have the right to say that there may be a science, too, of what is individual.

We are dealing with a science that is comprehensive because interpretative and non-comprehensive in default of interpretation. Comprehension could indeed only be successfully opposed to interpretation if the latter were simply compounded with observation. Now interpretation, on the contrary, differs from observation by the hypothesis which interpretation sets up for it and by the meaning or significance which (comprehensive in its turn) it imposes on observation. And when we have taken intelligible reduction as far as we can, what if we find the casuality was not to be identified and that there remains a residuum—

a variable that is irreducible or to become irreversible? Given such an eventual residuum, could anyone think that comprehension (even if it took the form of a 'clear apprehending by feeling' or intuition—such as we should like to put in place of interpretation) would really have anything better to offer? No, rather it is interpretation, because it has taken analysis almost to its limits, that can justifiably advance individual origins or derivation and confidently assign it as a cause. Interpretation and comprehension are not to be set up as opposites, nor yet should they be strangers. They are closely related in their nature and ought to exist together in amity.

I

PROFESSIONAL ETHICS

THE science of morals and rights should be based on the study of moral and juridical facts. These facts consist of rules of conduct that have received sanction. The problems to be solved in this field of study are :

(1) How these rules were established in the course of time: that is, what were the causes that gave rise to them and the useful ends they serve.

(2) The way in which they operate in society; that is, how they are applied by individuals.

Another matter, obviously, is to consider how we arrived at our current ideas of property and how theft has come to be a crime in certain conditions determined by the law; we must, too, define the conditions that account for the protective rule of the rights of property being here more and there less observed, that is, how it happens that some societies have more, and some, fewer thieves. These two questions are distinct, but even so, they could not be treated separately, for they are closely linked. There are the causes which have led to the establishment of rule, or law and order, and there are the causes responsible for the ascendancy of rule over the minds of men, sometimes over few, sometimes many. These causes are not identical but are yet of a kind to act as a check on each other and also to throw light one on the other. The problem of the origin and the problem of the operation of the function must therefore form the subject matter of research. This is why the equipment of the method used in studying the science of morals and rights is of two kinds. On the one hand we have comparative history and ethnography, which enable us to get at the origin of the rule, and show us its component elements first dissociated and then accumulating by degrees. In the second place there are comparative statistics, which allow us to compute the degree of relative authority with which this rule is clothed in individual consciousnesses and to discover the causes which make this authority variable. It is true we are not at present able to treat every moral problem from both points of view, for very often statistical data are lacking. This is perhaps the moment to remark that a science with its own technique ought to tackle both these questions.

In thus defining the subject of our inquiry, we have at the same time settled its sub-divisions. The moral and juridical facts—let us say, briefly, just moral facts—consist of rules of conduct that have sanction. Sanction is thus the feature common to all facts of this kind. No other kind of fact within the human order shows this peculiarity. For sanction, as we have defined it, is not simply any consequence following automatically on the act of a human being, as when we say, misusing the term, that intemperance brings illness as its sanction or laziness of the candidate, failure in examinations. Sanction is certainly a consequence of the act, but a consequence which results not from the act taken in isolation but from the conforming or not conforming to a rule of conduct already laid down. Theft is punished and this penalty is a sanction. But that is not because theft consists of this or that operation in the material sphere: the repressive counter-action that sanctions the right of property is entirely due to the fact that theft, that is, an assault on the property of another, is forbidden. Theft is punished only because it is prohibited. Let us suppose a society having a concept of property different from that which we hold: acts considered to-day as thefts and punished as such would then lose their significance and cease to be checked. The sanction, then, is not due to the essential nature of the act, since it can be withdrawn, the act remaining what it was. The sanction depends absolutely on the relation that exists between this act and a regulation governing its toleration or prohibition. This will explain why it is by reference to the sanction that all the rules of law and morality are defined.

That being so, the sanction, as the essential element of any moral rule at all, should naturally form the primary object of our inquiry. This is why the first part of these lectures has been devoted to a theory of sanctions. We have distinguished the different kinds of sanction—penal, moral, civil; and we have sought for the root common to them all and starting from that root, we have enquired how their differences came to be determined. This study of sanctions has been made independently of any consideration as to the rules themselves. Having thus isolated the features they have in common, we should turn to the rules. It is in them that the vital part—the heart—of this science lies.

Now these rules are of two kinds. The first apply to all men alike. They are those relating to mankind in general, that is, to each one of us as to our neighbour. All rules that set out the way in which men must be respected and their progress advanced—whether it be ourselves or our fellow-men—are equally valid for all mankind without exception. These rules of universal moral application are again divided into two groups: those concerning the relation of each one of us to his own self, that is, those that make up the moral code called 'individual'; and those concerning the relations we maintain with other people, with the exception of any particular grouping. The obligations laid upon us by both the one and the other arise solely from our intrinsic human nature or from the intrinsic human nature of those with whom we find ourselves in relation. They could not therefore vary from one individual to another, in the face of an identical moral consciousness. We have examined the first of these two groups of rules and the study of the second will form the latter part of the lectures. We should not, by the way, be surprised that these two divisions of morals, so closely related in some aspects, are so widely separated in our study and lie at the two extremes of the science. This division is quite reasonable. The function of the rules of the individual moral code is in fact to fix in the individual consciousness the seat of all morals—their foundations, in the widest sense: it is on these foundations that all else rests. On the other hand, the rules which determine the duties that men owe to their fellows, solely as other men, form the highest point in ethics, This, then, is the climax and it is the sublimation of all the rest. The order of the inquiry is thus not an artificial one: it corresponds exactly to the order of things.

But between these two extremes lie duties of a different kind. They depend not on our intrinsic human nature in general but on particular qualities not exhibited by all men. It has

3

been observed by Aristotle that, to some degree, morals vary according to the agents who practise them. The morals of a man, he said, are not those of a woman, and the morals of the adult not those of the child; those of the slave are not those of the master, and so on. . . . The observation is on the mark and it has nowadays a far greater field of application than Aristotle could have imagined. In reality, the greater part of our duties have this character. That indeed applied to those we had occasion to study last year, I mean to those duties which as a whole constitute the rights and the morals of the family. There indeed we find the difference of the sexes, of the ages, the difference that arises from a greater or lesser degree of kinship, and all these differences affect moral relations. It is the same, too, with the duties we shall be studying shortly, that is, civic duties, or those of man towards the State. For, since all men are not subject to the same State, they have by this fact duties which differ and are sometimes in opposition. Leaving aside entirely the antagonisms thus caused, civic obligations vary according to the State, and all States have not the same basis. The duties of the citizen are not the same in an aristocracy as in a democracy or in a democracy as in a monarchy. Family duties and civic duties do, however, exhibit a fairly large measure of common ground. For everyone, in principle, belongs to a family and founds one. Everyone is father, mother, uncle and so on. . . . And whilst everyone is not of the same age at the same moment, nor, therefore, has the same duties within the family, these differences are only fugitive, and whilst these various duties are not fulfilled at the same time by all, they are fulfilled by each one successively. There are no duties which man has not had to assume—at least, in the normal course of things. Only the differences based on sex endure and they tend to diminish to mere shades of difference. In the same way, whilst civic morals change according to the State, everyone nevertheless is the subject of a State and for that reason has duties, which everywhere have a similarity in their basic features—(duties of loyalty, service). No man exists who is not a citizen of a State. But there are rules of one kind where the diversity is far more marked; they are those which taken together constitute professional ethics. As professors, we have duties which are not those of merchants.

4

Those of the industrialist are quite different from those of the soldier, those of the soldier from those of the priest, and so on. . . . We might say in this connection that there are as many forms of morals as there are different callings, and since, in theory, each individual carries on only one calling, the result is that these different forms of morals apply to entirely different groups of individuals. These differences may even go so far as to present a clear contrast. Of these morals, not only is one kind distinct from the other, but between some kinds there is real opposition. The scientist has the duty of developing his critical sense, of submitting his judgment to no authority other than reason; he must school himself to have an open mind. The priest or the soldier, in some respects, have a wholly different duty. Passive obedience, within prescribed limits, may for them be obligatory. It is the doctor's duty on occasion to lie, or not to tell the truth he knows. A man of the other professions has a contrary duty. Here, then, we find within every society a plurality of morals that operate on parallel lines. It is with this part of ethics we shall be concerned. The place we assign to it in the course of this study is thus exactly in line with its features we have just identified. This moral particularism— if we may call it so—which has no place in individual morals, makes an appearance in the domestic morals of the family, goes on to reach its climax in professional ethics, to decline with civic morals and to pass away once more with the morals that govern the relations of men as human beings. In this respect, then, professional ethics find their right place between the family morals already mentioned and civic morals, that we shall speak of later. We shall therefore have a few words to say about professional ethics.

We shall only touch on them briefly, for it is obviously impossible to describe the code of morals proper to each calling and to expound them—their description alone would be a vast undertaking. It only remains to make a few comments on the more important aspects of the subject. We may reduce these to two: (1) what is the general nature of professional ethics compared with any other province of ethics? (2) what are the general conditions necessary for establishing any professional ethics and for their normal working?

The distinctive feature of this kind of morals and what

5

differentiates it from other branches of ethics, is the sort of uncon-
cern with which the public consciousness regards it. There are no
moral rules whose infringement, in general at least, is looked on
with so much indulgence by public opinion. The transgressions
which have only to do with the practice of the profession, come
in merely for a rather vague censure outside the strictly professional field. They count as venial. A penalty by way of discipline, for instance, imposed on a public servant by his official
superiors or the special tribunals to which he is responsible,
never sullies the good name of the culprit seriously, unless of
course it were at the same time an offence against common
morality.—A tax collector who commits some unscrupulous
action is treated as any other perpetrator of such actions ;
but a book-keeper who is complacent about the rules of
scrupulous accounting, or an official who as a rule lacks
energy in carrying out his duties, does not give the impression
of a guilty person, although he is treated as such in the organization to which he belongs. The fact of not honouring one's signa-
ture is a disgrace, almost the supremely shameful act, in
business. Elsewhere it is looked on with a very different eye.
We do not think of withholding respect from a bankrupt who
is only bankrupt. This feature of professional ethics can moreover easily be explained. They cannot be of deep concern to
the common consciousness precisely because they are not
common to all members of the society and because, to put it in
another way, they are rather outside the common consciousness. It is exactly because they govern functions not performed
by everyone, that not everyone is able to have a sense of what
these functions are, of what they ought to be, or of what special
relations should exist between the individuals concerned with
applying them. All this escapes public opinion in a greater or
lesser degree or is at least partly outside its immediate sphere
of action. This is why public sentiment is only mildly shocked
by transgression of this kind. This sentiment is stirred only by
transgressions so grave that they are likely to have wide general
repercussions.

It is this very fact which is a pointer to the fundamental
condition without which no professional ethics can exist. A
system of morals is always the affair of a group and can operate
only if this group protects them by its authority. It is made up

of rules which govern individuals, which compel them to act in such and such a way, and which impose limits to their inclinations and forbid them to go beyond. Now there is only one moral power—moral, and hence common to all—which stands above the individual and which can legitimately make laws for him, and that is collective power. To the extent the individual is left to his own devices and freed from all social constraint, he is unfettered too by all moral constraint. It is not possible for professional ethics to escape this fundamental condition of any system of morals. Since, then, the society as a whole feels no concern in professional ethics, it is imperative that there be special groups in the society, within which these morals may be evolved, and whose business it is to see they be observed. Such groups are and can only be formed by bringing together individuals of the same profession or professional groups. Furthermore, whilst common morality has the mass of society as its sole sub-stratum and only organ, the organs of professional ethics are manifold. There are as many of these as there are professions; each of these organs—in relation to one another as well as in relation to society as a whole—enjoys a comparative autonomy, since each is alone competent to deal with the relations it is appointed to regulate. And thus the peculiar characteristic of this kind of morals shows up with even greater point than any so far made: we see in it a real decentralization of the moral life. Whilst public opinion, which lies at the base of common morality, is diffused throughout society, without our being able to say exactly that it lies in one place rather than another, the ethics of each profession are localized within a limited region. Thus, centres of a moral life are formed which, although bound up together, are distinct, and the differentiation in function amounts to a kind of moral polymorphism.

From this proposition another follows at once by way of corollary. Each branch of professional ethics being the product of the professional group, its nature will be that of the group. In general, all things being equal, the greater the strength of the group structure, the more numerous are the moral rules appropriate to it and the greater the authority they have over their members. For the more closely the group coheres, the closer and more frequent the contact of the individuals, and,

7

the more frequent and intimate these contacts and the more exchange there is of ideas and sentiments, the more does a public opinion spread to cover a greater number of things. This is precisely because a greater number of things is placed at the disposal of all. Imagine, on the other hand, a population scattered over a vast area, without the different elements being able to communicate easily; each man would live for himself alone and public opinion would develop only in rare cases entailing a laborious calling together of these scattered sections. But when the group is strong, its authority communicates itself to the moral discipline it establishes and this, it follows, is respected to the same degree. On the other hand, a society lacking in stability, whose discipline it is easy to escape and whose existence is not always felt, can communicate only a very feeble influence to the precepts it lays down. Accordingly, it can be said that professional ethics will be the more developed, and the more advanced in their operation, the greater the stability and the better the organization of the professional groups themselves.

That condition is adequately fulfilled by a number of the professions. This applies above all to those more or less directly connected with the State, that is, those having a public character, such as the army, education, the Law, the government and so on. . . . Each one of these groups of functions forms a clearly defined body having its own unity and its own particular regulations, special agencies being instructed to see these are enforced. These agencies are sometimes officials appointed to supervise the work of their subordinates (inspectors, directors, seniors of all kinds in the official hierarchy). Sometimes they are regular tribunals, nominated by election or otherwise, and charged with preventing any serious defections from professional duty (supreme councils of the law, of public education, disciplinary boards of all kinds). Besides these callings there is one which is not of an official kind in the same degree but which has, however, an organization of a certain similarity: this is the advocates' association. The association (or 'order', to use the recognized term) is in fact an organized corporate body that holds regular meetings and is subject to an elected council, whose business it is to enforce the traditional rules applying to the group. In all these instances the cohesion of the

8

group is clearly seen and assured by its very organization. There is also to be found a pervading discipline that regulates all details of the functional activity and is capable of enforcing it if needs be.

Nevertheless—and this is the comment that matters most—there is a whole category of functions that do not satisfy this condition in any way: these are the economic functions, both industry and trade. Clearly, the individuals who follow the same calling are on terms with one another by the very fact of sharing a like occupation. Their very competition brings them in touch. But there is nothing steady about these connexions: they depend on chance meetings and concern only the individuals. It is a matter of this manufacturer being in touch with the other: it is not a matter of a body of members of one and the same industry meeting at fixed periods. What is more, there is no corporate body set above all the members of a profession to maintain some sort of unity, to serve as the repository of traditions and common practices and see they are observed at need. There is no organ of this description, because it can only be the expression of a life common to the group, and the group has no life in common—at least, not in any sustained kind of way. It is quite an exception to find a whole group of workers of this sort meeting in conference to deal with questions of general interest. These conferences last only for a time; they do not endure beyond the special occasion for which they were convened, and so the collective life they evoked dies with them.

Now, this lack of organization in the business professions has one consequence of the greatest moment: that is, that in this whole sphere of social life, no professional ethics exist. Or at least, if they do they are so rudimentary that at the very most one can see in them maybe a pattern and a foreshadowing for the future. Since by the force of circumstance there is some contact between the individuals, some ideas in common do indeed emerge and thus some precepts of conduct, but how vaguely and with how little authority. If we were to attempt to fix in definite language the ideas current on what the relations should be of the employee with his chief, of the workman with the manager, of the rival manufacturers with each other and with the public—what vague and equivocal formulas we should get! Some hazy generalizations on the loyalty and devotion

9

owed by staff and workmen to those employing them; some phrases on the moderation the employer should use in his economic dominance; some reproach for any too overtly unfair competition—that is about all there is in the moral consciousness of the various professions we are discussing. Injunctions as vague and as far removed from the facts as these could not have any very great effect on conduct. Moreover, there is nowhere any organ with the duty of seeing they are enforced. They have no sanctions other than those which a diffused public opinion has at hand, and since that opinion is not kept lively by frequent contact between individuals and since it therefore cannot exercise enough control over individual actions, it is lacking both in stability and authority. The result is that professional ethics weigh very lightly on the consciousnesses and are reduced to something so slight that they might as well not be. Thus, there exists to-day a whole range of collective activity outside the sphere of morals and which is almost entirely removed from the moderating effect of obligations.

Is this state of affairs a normal one? It has had the support of famous doctrines. To start with, there is the classical economic theory according to which the free play of economic agreements should adjust itself and reach stability automatically, without its being necessary or even possible to submit it to any restraining forces. This, in a sense, underlies most of the Socialist doctrines. Socialist theory, in fact, like classical economic theory, holds that economic life is equipped to organize itself and to function in an orderly way and in harmony, without any moral authority intervening; this, however, depends on a radical change in the laws of property, so that things cease to be in the exclusive ownership of individuals or families and instead, are transferred to the hands of the society. Once this were done, the State would do no more than keep accurate statistics of the wealth produced over given periods and distribute this wealth amongst the associate members according to an agreed formula. Now, both these theories do no more than raise a *de facto* state of affairs which is unhealthy, to the level of a *de jure* state of affairs. It is true, indeed, that economic life has this character at the present day, but it is impossible for it to preserve this, even at the price of a thoroughgoing change in the structure of property. It is not possible for a social function to exist without moral discipline. Otherwise, nothing remains but individual appetites, and since they are by nature boundless and insatiable, if there is nothing to control them they will not be able to control themselves.

And it is precisely due to this fact that the crisis has arisen from which the European societies are now suffering. For two centuries economic life has taken on an expansion it never knew before. From being a secondary function, despised and left to inferior classes, it passed on to one of first rank. We see the military, governmental and religious functions falling back more and more in face of it. The scientific functions alone are in a position to dispute its ground, and even science has hardly any prestige in the eyes of the present day, except in so far as it may serve what is materially useful, that is to say, serve for the most part the business professions. There has been talk, and not without reason, of societies becoming mainly industrial. A form of activity that promises to occupy such a place in society taken as a whole cannot be exempt from all precise moral regulation, without a state of anarchy ensuing. The forces thus released can have no guidance for their normal development, since there is nothing to point out where a halt should be called. There is a head-on clash when the moves of rivals conflict, as they attempt to encroach on another's field or to beat him down or drive him out. Certainly the stronger succeed in crushing the not so strong or at any rate in reducing them to a state of subjection. But since this subjection is only a *de facto* condition sanctioned by no kind of morals, it is accepted only under duress until the longed-for day of revenge. Peace treaties signed in this fashion are always provisional, forms of truce that do not mean peace to men's minds. This is how these ever-recurring conflicts arise between the different factions of the economic structure. If we put forward this anarchic competition as an ideal we should adhere to—one that should even be put into practice more radically than it is to-day—then we should be confusing sickness with a condition of good health. On the other hand, we shall not get away from this simply by modifying once and for all the lay-out of economic life; for whatever we contrive, whatever new arrangements be introduced, it will still not become other than it is or change its nature. By its very nature, it cannot be self-sufficing. A state of order or peace amongst men cannot follow of itself from any entirely material causes, from any blind mechanism, however scientific it may be. It is a moral task.

From yet another point of view, this amoral character of economic life amounts to a public danger. The functions of this order to-day absorb the energies of the greater part of the nation. The lives of a host of individuals are passed in the industrial and commercial sphere. Hence, it follows that, as those in this *milieu* have only a faint impress of morality, the greater part of their existence is passed divorced from any moral influence. How could such a state of affairs fail to be a source of demoralization? If a sense of duty is to take strong root in us, the very circumstances of our life must serve to keep it always active. There must be a group about us to call it to mind all the time and, as often happens, when we are tempted to turn a deaf ear. A way of behaviour, no matter what it be, is set on a steady course only through habit and exercise. If we live amorally for a good part of the day, how can we keep the springs of morality from going slack in us? We are not naturally inclined to put ourselves out or to use self-restraint; if we are not encouraged at every step to exercise the restraint upon which all morals depend, how should we get the habit of it? If we follow no rule except that of a clear self-interest, in the occupations that take up nearly the whole of our time, how should we acquire a taste for any disinterestedness, or selflessness or sacrifice? Let us see, then, how the unleashing of economic interests has been accompanied by a debasing of public morality. We find that the manufacturer, the merchant, the workman, the employee, in carrying on his occupation, is aware of no influence set above him to check his egotism; he is subject to no moral discipline whatever and so he scouts any discipline at all of this kind.

It is therefore extremely important that economic life should be regulated, should have its moral standards raised, so that the conflicts that disturb it have an end, and further, that individuals should cease to live thus within a moral vacuum where the life-blood drains away even from individual morality. For in this order of social functions there is need for professional ethics to be established, nearer the concrete, closer to the facts, with a wider scope than anything existing to-day.

There should be rules telling each of the workers his rights and his duties, not vaguely in general terms but in precise detail, having in view the most ordinary day-to-day occurrences. All these various inter-relations cannot remain for ever in a state of fluctuating balance. A system of ethics, however, is not to be improvised. It is the task of the very group to which they are to apply. When they fail, it is because the cohesion of the group is at fault, because as a group its existence is too shadowy and the rudimentary state of its ethics goes to show its lack of integration. Therefore, the true cure for the evil is to give the professional groups in the economic order a stability they so far do not possess. Whilst the craft union or corporate body is nowadays only a collection of individuals who have no lasting ties one with another, it must become or return to being a well-defined and organized association. Any notion of this kind, however, comes up against historical prejudices that make it still repugnant to most, and on that account it is necessary to dispel them.

II

PROFESSIONAL ETHICS (Continued)

THERE is no form of social activity which can do without the appropriate moral discipline. In fact, every social group, whether to be limited or of some size, is a whole made up of its parts: the primary element, whose repetition forms the whole, being the individual. Now, in order that such a group may persist, each part must operate, not as if it stood alone, that is, as if it were itself the whole; on the contrary, each part must behave in a way that enables the whole to survive. But the conditions of existence of the whole are not those of the part, by the very fact that these are two different things. The interests of the individual are not those of the group he belongs to and indeed there is often a real antagonism between the one and the other. These social interests that the individual has to take into account are only dimly perceived by him: sometimes he fails to perceive them at all, because they are exterior to himself and because they are the interests of something that is not himself. He is not constantly aware of them, as he is of all that concerns and interests himself. It seems, then, that there should be some system which brings them to mind, which obliges him to respect them, and this system can be no other than a moral discipline. For all discipline of this kind is a code of rules that lays down for the individual what he should do so as not to damage collective interests and so as not to disorganize the society of which he forms a part. If he allowed himself to follow his bent, there would be no reason why he should not make his way or, at very least, try to make his way, regardless of everyone in his path and without concern for any disturbance he might be causing about him. It is this discipline that curbs him, that marks the boundaries, that tells him what his relations with his associates should be, where illicit encroachments begin, and what he must pay in current dues towards the maintenance of the community. Since the precise function of this discipline is to confront the individual with aims that are not his own, that are beyond his grasp and exterior to him, the discipline seems to him—and in some ways is so in reality—as something exterior to himself and also dominating him. It is this transcendent nature of morals that finds expression in popular concepts when we find them turning the fundamental principles of ethics into a law deriving from a divine source. And the bigger the social group becomes, the more this making of rules becomes necessary. For, when the group is small, the individual and the society are not far apart; the whole is barely distinguishable from the part, and each individual can therefore discern the interests of the whole at first hand, along with the links that bind the interests of the whole to those of each one. But as the society expands, so does the difference become more marked. The individual can take in no more than a small stretch of the social horizon; thus, if the rules do not prescribe what he should do to make his actions conform to collective aims, it is inevitable that these aims will become anti-social.

For this reason, no professional activity can be without its own ethics. And, indeed, we have seen that very many of the professions do satisfy this desideratum. It is the functions of the economic order alone that form an exception. Even here, some rudiments of professional ethics are not lacking, but they are so little developed and have so little weight that they might as well not exist. This moral anarchy has been claimed, it is true, as a right of economic life. It is said that for normal usage there is no need of regulation. But from what source could it derive such a privilege? How should this particular social function be exempt from a condition which is the most fundamental to any social structure? Clearly, if there has been self-delusion to this degree amongst the classical economists it is because the economic functions were studied as if they were an end in themselves, without considering what further reaction they might have on the whole social order. Judged in this way, productive output seemed to be the sole primary aim in all industrial activity. In some ways it might appear that output, to be intensive, had no need at all to be regulated; that on the contrary, the best thing were to leave individual businesses and enterprises of self-interest to excite and spur on one another in hot competition, instead of our trying to curb and keep them within bounds. But production is not all, and if industry can only bring its output to this pitch by keeping up a chronic state of warfare and endless dissatisfaction amongst the producers, there is nothing to balance the evil it does. Even from the strictly utilitarian standpoint, what is the purpose of heaping up riches if they do not serve to abate the desires of the greatest number, but, on the contrary, only rouse their impatience for gain? That would be to lose sight of the fact that economic functions are not an end in themselves but only a means to an end; that they are one of the organs of social life and that social life is above all a harmonious community of endeavours, when minds and will come together to work for the same aim. Society has no justification if it does not bring a little peace to men—peace in their hearts and peace in their mutual inter-course. If, then, industry can be productive only by disturbing their peace and unleashing warfare, it is not worth the cost. In addition, even taking economic interests alone, a high rate of output is not everything. Value attaches to regularity as well. Not only is it essential that the article be produced in quantity, but that there be a regular flow of material sufficient to occupy the labour force. There should not be alternating periods of over and under production. No regulated planning means no regularity.

Classical economic theory has often boasted of the disappear-ance of former scarcities which in fact have become impossible since the lowering of tariffs and the ease of communications allow one country to get from others the supplies it happens to need. But the former crises in food supplies have given place to commercial and industrial crises that are no less scandalous in the disturbance they cause. And the more the dimensions of societies increase and the more the markets expand, the greater the urgency of some regulation to put an end to this instability. Because, as discussed earlier, the more the whole exceeds the part, the more the society extends beyond the individual, the less can the individual sense within himself the social needs and the social interests he is bound to take into account.

Now, if these professional ethics are to become established in the economic order, then the professional group, hardly to be found in this sphere of social life, must be formed or revived. For it is this group alone that can work out a scheme of rules. At this point, however, we come up against a historical prejudice. The name in history of this professional group is the guild[1] and this guild is held to have been bound up with our political *ancien régime* and therefore as not being able to survive it. It appears that to claim a corporate organization for industry and commerce would be retrograde and, in principle, such reversions are rightly considered as unhealthy phenomena.

There is, however, a primary fact which should put us on our guard against this reasoning, and that is the very long history of the guilds. If they only dated from the Middle Ages we might indeed believe that as they had come into existence with the political system of those days they had inevitably to pass away with it. But they have in fact an origin far more ancient. The craft guild has been with us from the time that crafts first began and industry ceased to be purely agricultural—that is, since there have been towns. In Rome, it certainly goes back to the pre-historic era. A tradition related by Plutarch and Pliny attributed its institution to the king Numa. "The most admirable foundation of this king", it says, "was the division he made of the people by crafts. The city was made up of two peoples, or rather, there were two distinct parts. . . . To do away with this main and serious cause of dissension, he divided the whole people into a number of bodies and this was done according to crafts. There were the flute-players, the goldsmiths, the carpenters and so on. . . ." (*Numa*, XVII). It is true this is only a legend, but it suffices to prove the ancient history of these *collegia* of craftsmen. However, under the kings as well as under the republic they had so obscure an existence that we have little knowledge of how they were organized. But as early as the time of Cicero their number had become considerable. "All classes of workers seem possessed by the desire to multiply the craft associations. Under the Empire, we see the domain of the guilds expanding to a degree that has perhaps never since been surpassed, if one takes the economic differences into account." (Waltzing). Then came a time when all categories of workers (and these various classes were many, since the division of labour was already far advanced) seemed to

[1] N. B. French *'corporation'*.

17

be forming into *collegia*. It was the same with those who lived by
trading. At this very moment, the *collegia* changed in character.
At first they were always private groups that the State controlled
by ordinance only from a distance. They then became regular
organs of public life. It was only by government authority
that they became established and they fulfilled genuine official
functions. The guilds for foodstuffs (butchers, bakers and so
on . . .), for instance, were responsible for the general supply
of provisions. It was the same with the other trades, although
to a less degree. Carrying thus a public office, the members of
these trade guilds, in exchange for the services they rendered,
had certain privileges, granted to them by successive emperors.
Little by little the official, in principle of small import, took the
upper hand and the guilds became regular cog-wheels of the
administration. But having fallen thus under tutelage, they then
became so crushed by responsibilities that they wished to re-
gain their independence. The State, now all-powerful, opposed
this by making the occupation and the obligations of law and
order involved in it, hereditary. No one could free himself
except by putting forward someone else to take his place. The
guilds endured thus in servitude until the end of the Roman
Empire.

Once the Empire had passed away, nothing survived of the
guilds but barely perceptible traces in the cities of Roman
origin in Gaul and Germany. Moreover, the civil wars that
ravaged Gaul and later the invasions, had destroyed trade and
industry. The craftsmen, who saw in the guilds the source of
these onerous responsibilities, without adequate profit to com-
pensate them, had taken advantage of this state of the country
to flee the cities and disperse through the countryside. Thus, as
happened later, in the eighteenth century, the life of the guilds
in the first century of our era became very nearly extinct. If a
contemporary critic had taken note of the situation, he would
probably have drawn the conclusion that the life of the guilds
had come to an end because they no longer had any intrinsic
purpose, if indeed they had ever had such purpose: he might
have taken any attempt to revive them as a retrograde step
destined to fail, for the reason that movements in history cannot be arrested. This is why the economists, at the end of last
century, on the pretext that the guilds of the *ancien régime* were

no longer equal to their rôle, thought themselves warranted in seeing them as mere survivals of the past, lacking any roots in the present and whose last traces should certainly be done away with. And yet the facts should provide a striking contradiction to such reasoning. In all European societies the guilds, after being in eclipse for a time, began a new existence. They were to reappear round about the eleventh and twelfth centuries. "The eleventh and twelfth centuries", says Levasseur, "seem to have been the period when craftsmen began to feel the need to unite and formed their first associations." From the thirteenth century onwards we see the guilds flourishing once more and developing steadily until the day when a new decadence begins for them. Do we not discern in this ancient past and this persistent survival a proof that they do not depend on some merely contingent or haphazard circumstance peculiar to a given political regime, but on wide and fundamental causes? They have been a necessity from the foundation of the City to the Empire at its zenith, and from the dawn of Christian societies to the French Revolution. That is probably because they respond to some need at once profound and lasting. The argument that explains their violent dissolution at the end of the last century as proof that they were no longer in harmony with the new conditions of collective existence can be refuted: is this not done by the very fact that, having passed from the scene that first time, they established themselves once more of their own volition and in a new form? Is not the need felt at the present time by all the great European societies to bring back the guild system to life a symptom on the contrary that this radical abolition was in itself an unhealthy phenomenon, and a sign that to-day reforms are called for that are different from those of Turgot and even the very opposite?

There is, however, one thing that, generally speaking, makes us sceptical as to the usefulness any such re-organization might have. If it is to serve its purpose it must be above all through its moral consequences, for each trade or craft association would have to become the focus of a moral life *sui generis*. Now, the records left to us of the former guilds and the very impression we get of the remnants that still survive, do not lead us to believe them equal to such a rôle. It seems as if they could fulfil only utilitarian functions, as if they could only serve the material interests of the profession. Were they to be set up again, it would simply mean replacing individual egotism by corporate egotism. We form an idea of the guilds (as they were in the final years of their most recent period) as taken up entirely with holding on jealously to their privileges and exclusive rights or even with increasing them. Now, it is not likely that absorption in affairs so narrowly professional could have any very favourable reaction on the morality of the corporate body or its members. Still, we must beware of applying to the whole corporative system something that may have been true of particular guilds at a given moment of their history. This defect is far from being inherent in all guild organization; in the Roman guilds, for instance, it is not to be found at all. The pursuit of utilitarian ends was of quite minor importance to them. "The craft guilds of the Romans", Waltzing says, "were a long way from having so definite a professional character as those of the Middle Ages; with them we find neither regulation of methods, nor the prescribing of apprenticeship nor exclusive rights; nor was it their aim to collect the necessary funds for developing an industry." The association certainly gave them greater power to safeguard their common interests should the need arise. This was but one of the useful results that came from it: it was not its main justification. What, then, were its main functions? In the first place, the guild was a religious *collegium*. Each had its own particular deity, its own ritual which, when the means were available, was celebrated in a specially dedicated temple. In the same way as each family had their *lar familiaris*, each city its *genius publicus*, each *collegium* had its tutelary god or *genius collegii*. This cult practised by the crafts always had its festivals and these festivals were celebrated with sacrifices. And there were always banquets held in common. It was not only to do honour to the god of the guild that the fellow members came together, but also on other occasions. For instance, at the feast of the *strenia*, "the Roman cabinet-makers and ivory workers gathered together in their *schola;* they received five *denarii*, cakes, dates, etc. . . . at the expense of the funds." The domestic festival of the *cara cognatio* or *caresta* (cherished kindred) was also celebrated; and just as presents were given in families on the First of January (?), so on this occasion a distribution of monies from the common fund was made within the *collegia*. There has been speculation whether the guild had a relief fund, and whether it regularly helped those members who were in need. Opinions on this point are divided. But these contentions lose colour and point by the fact that these distributions of money and provisions made at the festivals, these banquets . . .(? always held in common) in any case took the place of relief and may have stood for indirect aid. At any rate, those in want knew that at certain periods they could count on this disguised subsidy. As a corollary of this religious side, the Roman guild had too a funerary aspect. Its members were united in the ritual of their cult during their life-time just as the *gentiles* were, and like them they wished to sleep their last sleep together. All the guilds rich enough to do so had a *columbarium* in common where each of the members had a right to be buried. When the *collegium* had not the means to buy a funerary holding, it at least ensured for its members an honourable funeral ritual at the expense of the common fund. But the first was the more general. A common ritual, communal banquets, communal festivals, a communal cemetery—do we not find here the distinguishing features of the Roman domestic structure? "Every *collegium*", says Waltzing, "was one big family. The community of calling and interests replaced the ties of blood, and had not the associates too, like the family, their common ritual, their meals in common, their common sepulchre? We have seen that the religious and funerary ceremonies were like those of the family: and like these too, they celebrated the cherished kindred and the ritual of the dead." Elsewhere he says: "These frequent feasts were a powerful element in turning the *collegium* into a big family. No other term could better describe the nature of the ties that linked the associates and there are many signs proving that a great sense of brotherhood existed. The members looked on each other as brothers and sometimes used this name amongst themselves." The more usual expression was *sodales*. This very word indeed expresses a spiritual kinship implying close brotherhood. The patron and patroness of the *collegium* often took the style of father and mother. A proof of the devotion that the associates had for their *collegia* is seen in the legacies and gifts they made to them. There are, too, the memorial inscriptions, which read *pius in collegio*—'he was the devoted

son of his *collegium'*—just like the saying and indeed the in-
scription, *pius in suos*. According to Boissier, this family life was
indeed the chief aim of all the Roman guilds. "But in the case
of the artisan guilds", he says, "the associates met chiefly for
the pleasure of enjoying life together and for a sociableness not
so limited as in the family circle and yet not so diluted as in the
city : they wanted to be in the company of friends and in that
way make life freer and more agreeable."

The Christian societies did not in their framework follow
the pattern of the city, and the medieval guilds equally had
no very close likeness to the Roman guilds. Yet they too created
a moral environment for their members. "The guild", says
Levasseur, "united those of the same calling by close ties.
Very often it was set up in the parish or in some particular
chapel and placed itself under the protection of a saint who
became the patron of the whole community. . . . It was there
in the chapel that the associates would meet together and there
too would attend solemn Mass with all due ceremony; there-
after the associates would leave in a body to spend the rest of
the day in joyful feasting. Looked at in this way, the medieval
guilds resembled very closely those of the Roman era." . . . "In
order to meet all the expenses, the guild had to have a budget,
and this was provided. . . . A portion of the funds was set aside
for works of charity. . . . The cooks (of Paris) devoted one-third
of their fines to support the aged poor of their trade who had
fallen on evil days through lack of trade or by old age. . . .
Long afterwards, in the eighteenth century, there is still to be
found an entry in the goldsmiths' accounts of a free loan of
200 *livres* to a goldsmith who had been ruined." The respective
duties of masters and workmen were, too, set out in clear-cut
rules fixed for each particular trade. So, also, were the obligations of the masters to one another. Once engaged, a workman could not arbitrarily break his engagement. "The statutes
of all guilds are at one in forbidding the hiring of a hand who
has not finished his time and impose a heavy fine on the master
who offers the hire and the hand who accepts it." But the hand,
on his side, could not be discharged without a reason. In the
case of the burnishers, the reasons for the dismissal had to be
approved by ten of the hands and by the four wardens of the
craft. A regulation decided for each craft whether night work

was permitted or not. In case of prohibition, it was expressly forbidden to the master to make his hands keep watch by night. Other regulations were intended to guarantee professional integrity. All kinds of precautions were taken to prevent the merchant or craftsman from deceiving the customer or from giving the merchandise an appearance unwarranted by its true quality. "The butchers were forbidden to inflate the meat, to mix tallow with lard, to sell dog's flesh, and so on . . .; the weavers to make cloth of wool supplied by usurers, because this wool might be simply a pledge put down as security for a debt. . . . The cutlers were prohibited from making knife-handles covered with silk or wound with brass or tin wire, to obscure the plain wood beneath and so deceive an unwitting buyer, etc. . . ." It is true there came a time (eighteenth century) when these regulations became more vexatious than useful and were exploited to safeguard the masters' privileges rather than to protect the good name of the profession and the straight dealing of its members. However, there is no institution where deterioration does not set in at some point in its history. It may be that it fails to adapt itself in time to meet conditions of a new era. Or it may be that it develops in a one-sided direction. In that case too great a strain is put on its resources, with the result that it loses its aptitude to give the services for which it was responsible. This may be a reason for seeking to reform it, but not for declaring it for ever useless and doing away with it.

Nevertheless, the facts related show clearly that the professional group is by no means incapable of being in itself a moral sphere, since this was its character in the past. It is even obvious that this was its main rôle during the greater part of its history. At any rate, this is only a particular instance of a more general law. Within any political society, we get a number of individuals who share the same ideas and interests, sentiments and occupations, in which the rest of the population have no part. When that occurs, it is inevitable that these individuals are carried along by the current of their similarities, as if under an impulsion; they feel a mutual attraction, they seek out one another, they enter into relations with one another and form compacts and so, by degrees, become a limited group with recognizable features, within the general society. Now, once the group is formed, nothing can hinder an appropriate moral life from evolving, a life that will carry the mark of the special conditions that brought it into being. For it is not possible for men to live together and have constant dealings without getting a sense of this whole which they create by close association; they cannot help but adhere to this whole, be taken up with it and reckon it in their conduct. Now this adherence to some thing that goes beyond the individual, and to the interests of the group he belongs to, is the very source of all moral activity. That sense of this whole becomes acute, and then, as it is applied to affairs of communal life—the most ordinary as well as the most important—it is translated into formulas, some more defined than others. It is at this point we have a corpus of moral rules already well on the way to being founded.

If nothing abnormal occurs to disturb the natural course of things, all this is bound to come about. Moreover, it is well for the individual and for the society equally that it should be so. It is a good thing for the society when the moral activity thus released becomes socialised, that is, regulated. If left entirely to individuals, it can only be chaotic and dissipated in conflicts: the society cannot be shaken by so much internal strife without injury. Still, the activity is too far removed from the special interests that have to be regulated and from antagonisms that have to be calmed for it to serve as a restraining force, either direct or through the medium of any public office. That is why it is in the interest of the moral activity itself to allow the particular groups that fulfil these functions to be formed on these lines. At times, the activity even has to hasten their formation or make it easier. The individual, in the same way, finds decided advantage in taking shelter under the roof of a collectivity that ensures peace for him. For anarchy is painful to him also, on his own account. He too suffers from the everlasting wranglings and endless friction that occur when relations between an individual and his fellows are not subject to any regulative influence. It is not a good thing for a man to live like this on a war footing amongst his closest comrades and to entrench himself always as though in the midst of enemies. This sensation of hostility all about him and the nervous strain involved in resisting it, this ceaseless mistrust one of another— all this is a source of pain; for though we may like a fight, we also love the joys of peace; we might say that the more highly and the more profoundly men are socialized, that is to say, civilized—for the two are synonomous—the more those joys are prized. That is why, when individuals who share the same interests come together, their purpose is not simply to safeguard those interests or to secure their development in face of rival associations. It is, rather, just to associate, for the sole pleasure of mixing with their fellows and of no longer feeling lost in the midst of adversaries, as well as for the pleasure of communing together, that is, in short, of being able to lead their lives with the same moral aim.

It was in much the same way that the morals of the family were evolved. We have the fact that family life has been and still is a centre of morality and a school of loyalty, of selfless-ness and moral communing: the high standing which we accord to the family inclines us to find the explanation in certain attributes peculiar to the family and not to be found elsewhere. We like to think that consanguinity makes for unusually potent moral sympathies. Yet we saw last year that this con-sanguinity had by no means the exceptional force it has been given credit for. Those not related by consanguinity have been very numerous in families over a long period of time: the kinship called artificial was acquired extremely easily and had all the effects of the natural kinship. The family is thus not solely or essentially a consanguine group. It is a group of individuals who happen to have been brought together within the political society by an especially close community of ideas, sentiments and interests. Consanguinity did no doubt con-tribute to bring about this community, but it has been no more than one of the factors.

The physical surroundings, the community of economic interests and community of worship have been elements no less important. Still, we know the moral rôle which the family has played in the history of morals and the force of the moral life that was set up in the group that has evolved. And why should it be otherwise with a moral life springing from the pro-fessional group? Certainly, we might expect that that life would have less vigour in some ways, not because the com-ponent elements would be feebler in quality but because they would be less numerous. The family is a group embracing the whole sum of existence; nothing escapes it, everything finds an echo within it. It is the political society in miniature. On the other hand, only one specified part of existence comes directly within the province of the professional group, namely, the part concerned with the occupation. Again, we must not lose sight of the very large place in life taken up by the professions, as their functions become more specialized and as the field of each individual activity becomes more confined within the limits of the function it is responsible for.

This close parallel of the family and the professional group is proved in detail and directly confirmed by the facts in the case of the Roman guild. We have seen, surely, that the guild was a great family, that it was formed on the very model of the domestic society, with banquets, festivals, worship, burial, all in common. And here, precisely because we can observe the guild at the start of its evolution, we perceive more clearly than at other stages how it was constituted partly with moral ends in view. As long as industry was exclusively agricultural, it had its natural framework in the family and in the territorial group made up of contiguous families in the village. As a rule, as long as exchange or barter was little developed, the life of the husbandman did not draw him away from his home. He subsisted on what he produced. The family was at the same time a professional group. When did the guild first appear? With the crafts. This was in fact because the crafts could no longer keep their exclusively domestic character. For a man to live by a craft there had to be customers, so that account had to be taken of what other craftsmen of the same trade were doing; they had to compete with them and they had to get on with them. Thus, a new form of social activity was established that went beyond the compass of the family, without having any appropriate framework. If it was not to remain unorganized, it had itself to create one; a group of a new kind therefore had to be formed with this aim. But new social forms that are set up are always old forms more or less modified and partly changed for the worse. The family, then, was the pattern on which the new grouping that came into being was modelled, but it could of course only imitate the essential features, without reproducing them exactly. And so it happened that the budding guild was a sort of family. Whilst this grouping took the pattern of the family, it was, however, in the form of a social activity that

26

was freeing itself by degrees from the authority of the family. It was a breaking up of the attributes of the family.

In emphasizing this parallel, I do not, however, say that the guilds or associations of the future should or could have this domestic character. It is obvious that the more they evolve the more they must develop original characteristics and the further they should get away from the antecedent groups for which they are in part the substitutes. The medieval guild system, in its time, recalled only very remotely the domestic structure. With all the greater reason, it must be the same with the corporate associations needed to-day.

But then the question comes up of knowing what these guilds should be. Having seen why they are necessary we must consider what form they should assume to play their part in present conditions of collective existence. Difficult as the problem is, we shall attempt to say something about it.

III

PROFESSIONAL ETHICS (End)

BESIDES the historic prejudice we spoke of last time, there is a further fact that has led to the guild system being discredited: it is the revulsion that is generally aroused by the idea of economic control by rule. In our own minds we see all regulation of this sort as a kind of policing, maybe vexatious, maybe endurable, and possibly calling forth some outward reaction from individuals, but making no appeal to the mind and without any root in the consciousness. It appears like some vast set of workshop regulations, far-reaching and framed in general terms: those who have to submit to them may obey in practice if they must, but they could not really want to have them. Thus, the discipline laid down by an individual and imposed by him in military fashion on other individuals who in point of fact are not concerned in wanting them, is confused by us with a collective discipline to which the members of a group are committed. Such discipline can only be maintained if it rests on a state of public opinion and has its roots in morals; it is these morals that count. An established control by rule does no more, shall we say, than define them with greater precision and give them sanction. It translates into precepts ideas and sentiments felt by all, that is, a common adherence to the same objective. So it would be strangely mistaking its nature only to regard its outer aspect and grasp the letter of it alone. From such an angle, this control may indeed have the appearance of being orders that are simply obstructive and prevent individuals from doing what they like, and all in an interest not their own. It is therefore natural enough that they seek to rid themselves of this obstruction or reduce it to a minimum. But beneath the letter lies the spirit that animates it: there are the ties of all kinds binding the individual to the group he is part of and to all that concerns that group; there are all these social sentiments, all these collective aspirations, these traditions we hold to and respect, giving sense and life to the rule and lighting up the way in which it is applied by individuals. So it is a strangely superficial notion—this view of the classical economists, to whom all collective discipline is a kind of rather tyrannous militarisation. In reality, when it is normal and what it ought to be, it is something very different. It is at once the epitome and the governing condition of a whole life in common which individuals have no less at heart than their own lives. And when we wish to see the guilds reorganized on a pattern we will presently try to define, it is not simply to have new codes superimposed on those existing; it is mainly so that economic activity should be permeated by ideas and needs other than individual ideas and needs, in fine, so that it should be socialized. It is, too, with the aim that the professions should become so many moral *milieux* and that these (comprising always the various organs of industrial and commercial life) should constantly foster the morality of the professions. As to the rules, although necessary and inevitable, they are but the outward expression of these fundamental principles. It is not a matter of co-ordinating any changes outwardly and mechanically, but of bringing men's minds into mutual understanding.

Moreover, it is not on economic grounds that the guild or corporative system seems to me essential but for moral reasons. It is only through the corporative system that the moral standard of economic life can be raised. We can give some idea of the present situation by saying that the greater part of the social functions (and this greater part means to-day the economic—so wide is their range) are almost devoid of any moral influence, at any rate in what is their own field. To be sure, the rules of common morality apply to them, but they are rules made for a life in common and not for this specific kind of life. Further, they are rules governing those relations of the specific kind of life which are not peculiar to industry and commerce: they do not apply to the others. And why, indeed, in the case of those others, should there be no need to submit to a moral influence? What is to become of public morality if there is so little trace of the principle of duty in this whole sphere that is so important in the social life? There are professional ethics for the priest, the soldier, the lawyer, the magistrate, and so on. Why should there not be one for trade and industry? Why should there not be obligations of the employee towards the employer and vice versa; or of business men one towards the other, so as to lessen or regulate the competition they set up and to prevent it from turning into a conflict sometimes—as to-day—almost as cruel as actual warfare? All these rights and obligations cannot, however, be the same in all branches of industry: they have to vary according to the conditions in each. The obligations in the agricultural industry are not those obtaining in the unhealthy industries, nor of course do those in commerce correspond to those in what we call industry, and so on. A comparison may serve to let us realize where we stand on these points. In the human body all visceral functions are controlled by a particular part of the nervous system other than the brain: this consists of the sympathetic nerve and the vagus or pneumo-gastric nerves. Well, in our society, too, there is a brain which controls the function of inter-relationship; but the visceral functions, the functions of the vegetative life or what corresponds to them, are subject to no regulative action. Let us imagine what would happen to the functions of heart, lungs, stomach and so on, if they were free like this of all discipline. . . . Just such a spectacle is presented by nations where there are no regulative organs of economic life. To be sure, the social brain, that is, the State, tries hard to take their place and carry out their functions. But it is unfitted for it and its intervention, when not simply powerless, causes troubles of another kind.

This is why I believe that no reform has greater urgency. I will not say it would achieve everything, but it is the preliminary condition that makes all the others possible. Let us suppose that by a miracle the whole system of property is entirely transformed overnight and that on the collectivist formula the means of production are taken out of the hands of individuals and made over absolutely to collective ownership. All the problems around us that we are debating to-day will still persist in their entirety. There will always be an economic mechanism and various agencies to combine in making it work. The rights and obligations of these various agencies therefore have to be determined and in the different branches of industry at that. So a corpus of rules has to be laid down, fixing the

stint of work, the pay of the members of staff and their obligations to one another, towards the community, and so on. This
means, then, that we should still be faced with a blank page to
work on. Supposing the means—the machinery of labour—had
been taken out of these hands or those and placed in others, we
should still not know how the machinery worked or what the
economic life should be, nor what to do in the face of this
change in conditions. The state of anarchy would still persist;
for, let me repeat, this state of anarchy comes about not from
this machinery being in these hands and not in those, but because the activity deriving from it is not regulated. And it will
not be regulated, nor its moral standard raised, by any witch-
craft. This control by rule and raising of moral standards can
be established neither by the scientist in his study nor by the
statesman; it has to be the task of the groups concerned.
Since these groups do not exist at the present time, it is of the
greatest urgency that they be created. The other problems can
only be usefully tackled after that.

Taking this as granted, it remains to study the form the
corporative bodies should have if they are to be in harmony
with present-day conditions of our collective existence. Clearly,
there can be no question of restoring them in the form they had
in the past. They died out because they could no longer survive
as they were. But then, what is the form they are destined to
take? The problem is not an easy one. To solve it, we shall have
to be a bit methodical and objective, so we must first arrive
at how the guild system evolved in the past, and what the conditions were that set the evolution going. We might then judge
with some assurance what the system should become, given the
conditions at present obtaining in our societies. To do this,
however, further research is needed. Even so, it is not beyond
us to make out the general lines of development.

Although, as we have seen, the guild system goes back as far
as the early days of the Roman city, it was not in the age of
Rome what it became later on, in the Middle Ages. The
difference did not lie simply in the *collegia* of Roman craftsmen
having a character at once more religious and less vocational
than the medieval guilds. These two institutions differed in a
far more important feature. In Rome, the guild was an extra-
social institution, at least in origin. The historian who attempts

to analyse the political structure of the Romans, will not encounter anything on his way that gives him an inkling of the
existence of the guilds. They did not come within the Roman
constitution as recognized and distinct units. At no time, in the
electoral assemblies or in the army rallies, did the craftsmen
assemble by *collegia*. The *collegium* as such was never known to
take part in public life, either as a body or represented by
special agents. At the outside, the question might apply to
three or four *collegia*, which we can possibly identify with four
of the centuries[1] formed by Servius Tullius—*tignarii, aerarii*
(carpenters, copper-smiths), and the trumpeters and hornblowers. But this is only conjectural. Very likely the centuries
thus classified did not take in all the carpenters, smiths and so
on, but only those who made or repaired arms and war equipment. Dionysius of Halicarnassus tells us explicitly that workers
grouped in this way had a solely military function $\epsilon \grave{\iota}s$ $\tau \grave{o} \nu$
$\pi \acute{o} \lambda \epsilon \mu o \nu$ and that in addition there were other workers grouped
under the same heading who in time of war had to perform
duties of another kind. We may therefore believe that these
centuries represented not the *collegia* but military sub-divisions.
In any case, as far as all the other *collegia* are concerned, they
were certainly outside the administrative structure of the
Roman people. Thus, these *collegia* were supererogatory: they
were as social forms more or less irregular or at least they could
not be reckoned as amongst those that were regular. This is
easy to understand. They were set up at a time when the
crafts were moving towards a certain development. Over a
long period the crafts were no more than a minor and subsidiary feature of collective activity in the Roman world.
Rome was essentially an agricultural and military society.
As an agricultural society it was divided into *gentes, curiae* and
tribes. Assembly by centuries[2] reflects rather the military side.
But it was quite natural that the industrial functions, at first
unknown, then only very rudimentary, should not affect the
political structure of the City in any way. They were *cadres* set
up late in the day alongside normal official *cadres*: the product
of a kind of outgrowth from the very early social structure of
Rome. Moreover, until a very late date in Roman history, the

---

[1] Meaning here 'infantry troops'.
[2] Meaning here 'voting unit in the assembly'.

craft carried the mark of a moral obloquy; that fact puts out of court any idea that it ever held an official place in the State. Things did no doubt change with time, but the very way in which they changed clearly demonstrates what they were like at the outset. The craftsmen had to have recourse to irregular means to see that their interests were respected and to secure a status in keeping with their growing importance. The *collegia* had to proceed by way of plotting and underground agitation. This is the surest evidence that the Roman society, in the ordinary way, was not open to them. And although later they ended by being integrated within the State, becoming cogs in the administrative machine, this position was no proud victory for them, nor profitable, but a grievous dependence. They did then gain entry into the State but not to occupy the place to which, it might seem, their services entitled them. It was simply so that they might be the more closely supervised and contolled by the governing authority. "The guild", says Levasseur, "became the chain that held them captive and which the Imperial hand tightened, the more arduous or the more necessary to the State their labour became. . . ." To sum up, the guilds, after having been kept outside the normal *cadres* of the Roman society, were in the end admitted but only to be reduced to a kind of servitude.

Their position in the Middle Ages was quite otherwise. From the outset, as soon as the guilds come on the scene, they give an impression of being the normal framework of that section of the population which was called upon to play a very considerable part in the State: this was the third estate, the commonalty or bourgeoisie. Indeed, for a long time the bourgeois and the craftsman were one and the same. "The bourgeoisie in the thirteenth century", says Levasseur, "was made up exclusively of craftsmen. The class of magistrates and jurists had hardly begun to take shape; the scholars still belonged to the clergy; the number of small freeholders (*rentiers*[1]) was very limited, because landed property was at that time almost wholly in the hands of the nobles; there remained to the commonalty only the labour of the workshop or the counting-house, and it was by their industry or trading that they had won a status in the kingdom." It was the same in Germany.

[1] Tr. note—*rentiers :* paying rent in money, kind or services.   (H.W.C. Davis)

33

The bourgeoisie was the population of the towns; now, we know that the towns in Germany had become established around permanent market sites set up by a feudal lord at some point of his domain. The population which had settled around these markets and which became the urban population was made up mainly of craftsmen and merchants. The towns from the beginning were the centres of manufacturing and trading activity: it is this fact that distinguishes the urban groups of Christian societies from those which are their counterpart— or appear to be—in other societies. The identity of both kinds of population was such that the terms *mercatores* and *forenses* are synonomous with that of *cives:* the same applies to *jus civilis* and *jus fori.* Thus, the framework of the crafts was the earliest structural form of the European bourgeoisie.

Likewise, when the towns, which in the beginning were seignorial dependencies, became free, and the *communes* were formed, the corporate body or craft guild, which had anticipated this transition, became the basis of the constitution of the *commune.* Indeed, "in almost all the *communes,* the political system and the election of magistrates were based on the division of citizens into craft guilds." Very often the voting was done through the craft guild and the heads of the corporate body and those of the *commune* were chosen at the same time. "At Amiens, for instance, the craftsmen met every year to elect the mayors of each corporation (guild) or banner; the elected mayors then nominated twelve *échevins* who brought in twelve others, and this corps of *échevins* in turn presented to the mayors of the banners three persons from whom they chose the mayor of the *commune.* . . . In some cities, the method of election was still more complex, but in all of them the political and municipal structure was closely linked with the structure of labour." And just as the *commune* was an aggregate of the craft guild, so the craft guild was a *commune* on a small scale. The guild had indeed been the model for the institution of the *commune,* which was a larger and more expanded form of it.

Let us sum up briefly. From being at first obscure, despised and exterior to the political constitution, we see the guild become the basic element in the *commune.* We know, on the other hand, what the *commune* has been in the history of all the great European societies: it became with time their corner-stone.

The *commune* is an aggregate of the guilds or corporate bodies and is itself formed on the guild model. From these facts we see that it is the guild, in the final analysis, that has served as a basis for the whole political system which emerged from the progress of the *commune*. In Rome, it was outside any *cadre* but of our own societies it was the basic framework or *cadre* itself. We see that, in its course, it has grown in dignity and significance to a remarkable degree. And there is still another reason for discrediting the hypothesis according to which it is destined to pass away. As we go on in history to the sixteenth and seventeenth centuries, the guild becomes a still more necessary element in the political structure. So there is small likelihood that of a sudden all justification for its existence should be lost. All to the contrary, it would be far more valid to hold that it will be called on to play an even more vital part in the future than in the past.

At the same time, the points just discussed enable us to discern, first, why decay set in about two centuries ago—that is, what has prevented the guild from being equal to the duties incumbent on it—and secondly, what its development should be to reach that level. We have seen that the guild, in its medieval form, was closely bound up with the whole structure of the *commune*. The two institutions were inter-related. Now, there was nothing to impair this solidarity as long as the crafts themselves had a communal character. As long as every craftsman and every merchant as a rule had as customers only those who lived in the same town or those who came in from the outskirts on market day, the craft guild with its closely localized structure met all needs. But it was a different matter when large-scale industry came in. Given its nature, it could not fit into the *cadres* of a town. For one thing, its site was not necessarily in a town: it could be set up at any point in the area, in the country as usefully as in the town—at all events, away from built-up areas; in fact wherever it could get supplies at the lowest economic cost and whence it could branch out furthest and most easily. Further, regular customers were secured all over the place and the sales range was confined to no particular region. An institution as closely involved in the *commune* as the guild was, could therefore not be of use in framing and regulating a form of social activity that was so completely

35

independent of the *commune*. Indeed, as soon as it came on the scene, large-scale industry found itself outside the old guild system. It was not however, for all that, free of any kind of control by regulation. It was the State that stood direct to industry as in earlier times the trade or craft guild stood to the urban trades. The royal authority granted privileges to the manufacturer with one hand and subjected him to its control with the other. Hence the title of 'Royal Manufactories' bestowed on them. This direct tutelage by the State was of course only feasible whilst manufactures were still few and in the early stages. The ancient guild in its early form failed to adapt itself to the new style of industry and the State was able to provide a substitute for the old guild discipline only for a period. It does not follow, even so, that all discipline henceforth was to serve no purpose, but merely that the earlier guild had to be reconstructed to operate in the new conditions of economic life. The change that had come about meant that industry, instead of being local and municipal, had become an affair of the whole country. From all of this we have to draw the conclusion that the guild, too, had to change in parallel fashion and in place of remaining a municipal institution, it had to become a public institution. Experience in the seventeenth and eighteenth centuries goes to prove that the guild system, which kept the pattern of a municipal affair, could not be appropriate to industries that in their wide scope and importance made their mark on the common interests of the society. On the other hand, that experience demonstrates that the State was itself not able to perform this office, because economic life is too vast and too complex, with too many ramifications, for it to supervise and regulate its operations effectively. Is not the lesson to be drawn from these facts, that the guild should assume a different character, and that it should get closer to the State without being absorbed by it? In fact, that it should become something national, whilst remaining a subsidiary group and relatively autonomous? The guild was too slow in transforming itself: it failed to bend before the pressure of new needs and so was broken. As it could not adapt itself to this new kind of life emerging, that life quite naturally receded from it. These are the facts that explain what the craft guild had become on the eve of the Revolution: a kind of dead substance

or foreign body which only persisted in our social organism by
the force of inertia. The moment had to come when it was
violently ejected. But the problem of the needs which the guild
could not satisfy was not solved by any such root and branch
abolition. And so we are left with this whole question, made
only more critical and more acute by a hundred years of fumbl-
ing and of distressing experiments. It does not, however, seem
impossible to solve.

Let us imagine—spread over the whole country—the various
industries grouped in separate categories based on similarity
and natural affinity. An administrative council, a kind of
miniature parliament, nominated by election, would preside
over each group. We go on to imagine this council or parliament as having the power, on a scale to be fixed, to regulate
whatever concerns the business: relations of employers and
employed—conditions of labour—wages and salaries—relations of competitors one with another, and so on . . . and there
we have the guild restored, but in an entirely novel form. The
establishment of this central organ appointed for the management of the group in general, would in no way exclude the
forming of subsidiary and regional organs under its direction
and subordinate to it. The general rules to be laid down by it
might be made specific and adapted to apply to various parts
of the area by industrial boards. These would be more regional
in character just as to-day under Parliament there are councils
for the *département* or municipality. In this way, economic life
would be organized, regulated and defined, without losing any
of its diversity. Such organization would do no more than
introduce into the economic order the reforms already made in
all other spheres of the national life. Customs, morals, political
administration, all of which formerly had a local character
and varied from place to place, have gradually moved towards
uniformity and to a loss of diversity. The former autonomous
organs, the tribunals, the feudal and communal powers, have
become with time auxiliary organs, subordinate to the central
organism that took shape. Is it not to be expected that the
economic order will be transformed with the same trend and by
the same process? What existed at the outset was a local structure, an affair of the community: what has to take its place is
not a complete absence of organization, a state of anarchy;

rather it would be a structure that was comprehensive and national, uniform and at the same time complex, in which the local groupings of the past would still survive, but simply as agencies to ensure communication and diversity.

It follows that the guild system would in this way be saved from another flaw that it was reproached with in the past, and rightly—that of being static. As long as its horizon was bounded by the walls of the city, it was inevitable that the guild should easily become the prisoner of tradition, like the city itself. In a group so hedged about, the conditions of life cannot change very much; habit has thus dominion over people and over things without any counter-balance and innovations in the end come even to be dreaded. The traditionalism of the guilds and their tendency to routine only reflected the prevailing traditionalism and had the same *raison d'être*. Still, it did outlive the causes from which it sprang and which were its original justification. The unification of the country, leading to the emergence of large-scale industry, resulted in a widening of perspectives and so to the awakening of a man's consciousness to new wants as to new ideas. He began to have aspirations hitherto unknown, a greater need of amenities and ease in living. Also, his tastes began to be more subject to change. It was otherwise with the guilds; they failed to change with the times or to be pliable; they kept rigidly to the old ways and customs and were in no state to respond to the new calls on them. Here we see another cause of the guilds' losing goodwill. But national corporate bodies would not be open to this danger. Their scope and their complexity would protect them against inertness. They would comprise elements that were too many and too diverse for a fixed uniformity to be feared. The equilibrium of such organization can be only relatively stable and would therefore be in complete harmony with the moral equilibrium of a society with the same character and in nowise rigid. Too many different minds would be at work within them for new re-arrangements not to be constantly preparing or, as it were, in a latent state. A group that extends over vast areas (such, for example, as China) is never static because change there is unceasing.

This seems to be the fundamental principle of the only kind of corporative system that would be appropriate to large-scale industry. We have shown the outlines, and it remains to solve a number of secondary questions that cannot be dealt with here. I shall only touch on the most important.

To begin with, it is often asked whether the guild should be compulsory, whether or no individuals should be bound to membership. This question, I feel, is only of limited interest. In fact, from the day when the guild system was set up, it would be such a handicap for the individual to remain aloof that he would join of his own accord, without any need of coercion. Once constituted, a collective force draws into its orbit those who are unattached: any who remain outside are unable to hold their ground. Moreover, it is beyond me to understand the scruples that some feel in this case against any suggestion of compulsion. Every citizen nowadays is obliged to be attached to a *commune* (parish). Why then should the same principle not apply to the profession or calling? All the more, since in fact the reform we are discussing would in the end result in the professional association taking the place of the jurisdictional area as a political unit of the region.

A more important matter is to know what the respective place and part of employer and employed would be in the corporative structure. It seems to me obvious that both should be represented in the governing body responsible for supervising the general affairs and well-being of the association. Such a body could only carry out its function provided that it included both these elements. However, one is forced to wonder whether a distinction would not have to be made at the base of the structure: whether the two categories of industrial personnel would not have to nominate their representatives separately—in a word, whether the electoral bodies would not have to be independent, at all events when their respective interests were obviously in conflict.

Finally, it seems certain that this whole framework should be attached to the central organ, that is, to the State. Occupational legislation could hardly be other than an application in particular of the law in general, just as professional ethics can only be a special form of common morality. To be sure, there will always be all the various forms of economic activity of individuals, which involve such overall regulation, and this cannot be the task of any group in particular.

So far, we have only briefly indicated the functions which might take shape in the corporative body. We cannot foresee all those which might be assigned to it in the future. Our best course is to keep to those which could be handed over to it straight away. From the legislative point of view, certain functions have to be classified according to the industry, such as the general principles of the labour contract, of salary and wages remuneration, of industrial health, of all that concerns the labour of women and children, etc., and the State is incapable of such classification. The provision of superannuation and provident funds, etc. cannot be made over without danger[1] to the funds of the State, overburdened as it is with various services, as well as being too far removed from the individual. Finally, the regulation of labour disputes, which cannot be codified as laws on any hard and fast principle, calls for special tribunals. In order to adjudicate with entire independence, these would have rights that varied with the varying forms of industry. There we have the judicial task, which might be assigned henceforth to the guilds in their revived and altered form. This threefold task would have to be assigned to these three (? ?[2]) organs or groups of organs: it is there you have practical problems that only experience would settle. The main thing is to set up the group and to give it a *raison d'être* by endowing it, very cautiously, with some of the functions just mentioned. Once it had been formed and had begun its life, it would develop of its own accord and no one can foresee at what point this evolution would stop. As I said earlier, the other reforms could only be tackled effectively when this first step had been taken: further, it is even possible that they might come about naturally from that step. If some re-casting of the laws of property is to come about, it is not the (? ?[2]) who can say for his part what form this will take. Anyone knowing the complexity of social life and the room it leaves for the play of the most conflicting elements, is aware of the over-simplification in formulas now current. It is hardly likely that the day will come when the means of production will be logically divorced from the means of consumption, when

[1] 'réservé sans danger' : should perhaps read : 'remis sans danger entre les mains de l'Etat '.
[2] Question marks represent gaps in the text, originally intended as lecture notes.

nothing will remain of the old rights of property, when the position of employer will no longer exist, and when all rights of inheritance will have been abolished. It is not within human foresight to say what part these facts of any future structure ... (omission), what portion of the past will permanently survive, and what . . . (omission) in the future . . .(omission).

This re-distribution can only come about of its own impetus, by the pressure of facts and experience. If industrial life be organized, that is, if it be given the organ it has need of, then this system, by coming in contact with other social organs, will of itself become a source of radical changes beyond our powers of imagining. Not only is the guild system . . .(? ? [1]).

---

[1] Question marks represent gaps in the text, originally intended as lecture notes.

IV

CIVIC MORALS

DEFINITION OF THE STATE

W E have studied in succession the moral and juridical rules that apply to the relation of the individual to himself, to the family group and to the professional group. We now have to set about examining the individual in his relations with another group, one greater in scope than the others, greater indeed than any other organized group in existence to-day, that is, the political group. The rules taken as a whole that have received sanction and that determine what these relations should be, form what is called civic morals. Before we begin this study, it is important to define what we understand by a political society.

An essential element that enters into the notion of any political group is the opposition between governing and governed, between authority and those subject to it. It is quite possible that in the beginning of social evolution this gap may not have existed; such an hypothesis is all the more likely since we do find societies in which the distance between the two is only faintly perceptible. But in any case, the societies where it is seen cannot be mistaken for those where it does not occur. The former differ from the latter in kind and require different terms of description: we should keep the word 'political' for the first category. For if this expression has any one meaning, it is, above all, organization, at any rate rudimentary; it is established authority (whether stable or intermittent, weak or strong), to whose action individuals are subject, whatever it be.

But an authority of this type is not found solely in political societies. The family has a head whose powers are sometimes limited by those of a family council. The patriarchal family of the Romans has often been compared to a State in miniature. Although, as we shall soon see, this expression is not justified, we could not quarrel with it if the sole distinguishing feature of the political society were a governmental structure. So we must look for some further characteristic.

This lies possibly in the especially close ties that bind any political society to its soil. There is said to be an enduring relationship between any nation and a given territory. "The State", says Bluntschli, "must have its domain; the nation demands a country." But the family, at least in many countries, is no less bound to the soil—that is, to some charted area. The family, too, has its domain from which it is inseparable, since that domain is inalienable. We have seen that the patrimony of landed estate was sometimes the very kernel of the family; it is this patrimony that made its unity and continuity and it was about about this focus that domestic life revolved. Nowhere, in any political society, has political territory had a status to compare with this in importance. We may add, however, that where cardinal importance attaches to national territory, it is of comparatively recent date. To begin with, it seems rather arbitrary to deny any political character to the great nomad societies whose structure was sometimes very elaborate. Again, in the past it was the number of citizens and not the territory that was considered to be the primary element of the State. To annex a State was not to annex the country but its inhabitants and to incorporate them within the annexing State. On the other hand, we may see the victors preparing to settle down in the country vanquished, without thereby losing their own cohesion or their political identity. During the whole early period of our history, the capital, that is, the territorial centre of gravity of the society, had an extreme mobility. It is not a great while since the peoples became so identified with the territories they inhabit, that is, with what we should call the geographical expression of those peoples. To-day, France is not only a mass of people consisting in the main of individuals speaking a certain language and who observe certain laws and so on, but essentially a certain defined part of Europe. If indeed all the Alsatians had opted for French nationality in 1870, we might have with justice still considered France as mutilated or diminished, by the sole fact that she had abandoned a delimited part of her soil to a foreign Power. But this identification of the society with its territory has only come about in those societies that are the most advanced. To be sure, it is due to many causes, to the higher social value that the soil has gained, perhaps also to the relatively greater importance that the geographical bond has assumed since other social ties of a more moral kind have lost their force. The society of which we are members is in our minds all the more a well-defined territory, since it is no longer in its essence a religion, a corpus of traditions peculiar to it or the cult of a particular dynasty.

Leaving territory aside, should we not find a feature of a political society in the numerical importance of the population? It is true we should not ordinarily give this name to social groups comprising a very small number of individuals. Even so, a dividing line of this kind would be extremely fluctuating: for at what precise moment does a concentration of people become of a size to be classified as a political group? According to Rousseau, it would be at the ten thousand figure, but Bluntschli rates this as too low. The estimates of both are equally arbitrary. A French *département* sometimes has more inhabitants than many of the City States of Greece and Italy. Any one of these, however, constitutes a State, whilst a *département* has no claim to such a term.

Nevertheless, we touch here on a distinctive feature. To be sure, we cannot say that a political society differs from family groups or from professional groups on the score that it has greater numbers, for the numerical strength of families may in some instances be considerable while the numerical strength of a State may be very small. But it remains true that there is no political society which does not comprise numerous different families or professional groups or both at once. If it were confined to a domestic society or family, it would be identical with it and hence be a domestic society. But the moment it is made up of a certain number of domestic societies, the resulting aggregate is something other than each of its elements. It is something new, which has to be described by a different word. Likewise, the political society cannot be identified with any professional group or with any caste, if caste there be; but is always an aggregate of various professions or various castes, as it is of different families. More often, when we get a society made up of a collection of secondary groups varying in kind, without itself being a secondary group in relation to a far bigger

44

society, then it constitutes a social entity of a specific kind. We should then define the political society as one formed by the coming together of a rather large number of secondary social groups, subject to the same one authority which is not itself subject to any other superior authority duly constituted.

Thus, and it should be noted, political societies are in part distinguished by the existence of secondary groups. Montesquieu was conscious of this in his day, in speaking of the social form that seemed to him the most highly organized, that is, the monarchy. He said that it involved "intermediary, subordinate and dependent powers." (*De l'Esprit des Lois*, Bk. II, ch. IV.) We can see the whole importance of these secondary groups we have been discussing so far. They are not only necessary for directing the particular interests, domestic or professional, that they include and that are their own *raison d'être;* they also form the primary condition for any higher organization. Far from being in opposition to the social group endowed with sovereign powers and called more specifically the State, the State presupposes their existence: it exists only where they exist. No secondary groups, no political authority—at least, no authority that this term can apply to without being inappropriate. Later on, we shall see the source of this solidarity that unites the two kinds of grouping. For the moment, it is enough to record the fact.

It is true that this definition runs counter to a theory long accepted as established: this is the theory to which Sumner Maine and Fustel de Coulanges have given their name. According to these authorities, the elementary society, from which the more composite societies are held to have sprung, is considered to be an extensive family group made up of all the individuals linked by ties of blood or ties of adoption and placed under the direction of the oldest male ascendant, the patriarch. This is the patriarchal theory. If this were a fact, we should find a constituted authority in the very beginning, analogous at all points with the authority we find in the more complex State; it would therefore be truly political, when in reality the society of which it is the key-stone is single and uncompounded, and not made up of any smaller societies. The supreme authority of cities, of kingdoms, of nations, constituted later on, would have no original and specific character whatever; it would derive from the patriarchal authority and be formed on its model. The society called political would be only families on a greater scale.

But this patriarchal theory is no longer tenable to-day; it is a hypothesis which rests on no fact whatever of direct observation, and which is disproved by a host of known facts. The patriarchal family as described by Sumner Maine and Fustel de Coulanges has never been under observation. A group made up of consanguines, living in a state of autonomy under the control of a more or less powerful head, has never been known. All the family groups that we do know which show even a vestige of organization and which recognize some definite authority, form part of greater societies. We define the clan as being at the same time a political and family sub-division of a wider social aggregate. But, it will be asked, how about the beginning? We may legitimately suppose that in the beginning there existed simple forms of society which did not comprise any society of a still simpler form; both logic and the analogies compel us to make a hypothesis which is confirmed by certain facts. On the other hand, nothing entitles us to think that such societies were subject to an authority of any kind. And one fact that should make us reject this hypothesis as altogether unlikely is that the more the clans of a tribe are independent one of another and the more each one tends towards autonomy, the more we look in vain for anything resembling an authority or any kind of governmental power. They are masses that are almost entirely amorphous or without structure, all the members of which are on the same level. Therefore the organization of partial groups, of clans, families and so on . . . did not precede the organization of the total aggregate which came about from their combination. We should not, however, go on to conclude that, conversely, the organization of the groups, etc. sprang from the organization of the aggregate. The truth is that they are interdependent, as we said just now, and that they condition each other mutually. The parts were not organized in the first instance to form a whole which was subsequently designed on their pattern, but the whole and the parts were organized at the same time. What also follows from the foregoing is that the political societies imply the existence of an authority: since this authority can only emerge where

46

the societies comprise within themselves a number of elementary societies, the political societies are of necessity polycellular or polysegmental. This is not to say that there have never been societies consisting of one segment alone, but they form a different species and are not political.

It remains true, however, that one and the same society may be political in some respects, and only constitute a partial and secondary group in others. This is what occurs in all federal States. Each individual State is autonomous to a certain degree: this degree is more limited than if there were not a federation with a regular structure, but the degree, although diminished by this federation, is not reduced to nil. Each member constitutes a political society, a State in the true meaning of the term, to the extent to which it is answerable only to itself and is not dependent on the central authority of the federation. On the other hand, to the extent to which it is subordinate to some organ superior to itself, it is an ordinary secondary group, a partial one and analogous to a district, a province, a clan or a caste. It ceases to be a whole and no longer emerges except as a part. Thus our definition does not establish an absolute line of demarcation between political societies and others; but that is because there is not and could not be such a line. On the contrary, the sequence of things is continuous. The major political societies are formed by the gradual aggregation of the minor. There are periods of transition when these minor societies, still keeping something of their original nature, begin to develop into something different and take on new characteristics, and when consequently, their status is ambiguous. The main thing is, not to record a break in continuity where none exists, but to be aware of the specific features which distinguish political societies and which (according to their degree of 'more or less') determine whether these societies are really more, or less, entitled to this term.

Now that we know the distinguishing marks of a political society, let us see what the morals are that relate to it. From the very definition just made, it follows that the essential rules of these morals are those determining the relation of individuals to this sovereign authority, to whose control they are subject. Since we need a word to indicate the particular group of officials entrusted with representing this authority, we are agreed to keep for this purpose the word 'State'. It is true that very often we apply the word State not to the instrument of government but to the political society as a whole, or to the people governed and its government taken as one, and we ourselves often use the term in this sense. It is in this way that we speak of the European States or that we call France a State. But since it is well to have separate terms for existent things as different as the society and one of its organs, we apply the term 'State' more especially to the agents of the sovereign authority, and 'political society' to the complex group of which the State is the highest organ. This being granted, the principal duties under civic morals are obviously those the citizen has towards the State and, conversely, those the State owes to the individual. To understand what these duties are, we must first of all determine the nature and function of the State.

It is true it may seem that we have already answered the first question and that the nature of the State has been defined at the same time as the political society. Is not the State the supreme authority to which the political society as a whole is subordinate? But in fact this term authority is pretty vague and needs definition. Where does the group of officials vested with this authority begin and end, and who constitute, properly speaking, the State? The question is all the more called for, since current speech creates much confusion on the subject. Every day, we hear that public services are State services; the Law, the army, the Church—where there is a national Church —are held to form part of the State. But we must not confuse with the State itself the secondary organs in the immediate field of its control, which in relation to it are only executive. At very least, the groups or special groups (for the State is complex)—to which these secondary groups (called more specifically administrative) are subordinate, must be distinguished from the State. The characteristic feature of the special groups is that they alone are entitled to think and to act instead of representing the society. The representations,[1] like the solutions that are worked out in this special *milieu* are inherently and of necessity collective. It is true, there are many representations and many collective decisions beyond those that take shape in this way. In every society there are or have

[1] NB. in E.D.'s sense of word.

been myths and dogmas, whenever the political society and the Church are one and the same, as well as historical and moral traditions: these make the representations common to all members of the society but are not in the special province of any one particular organ. There exist too at all times social currents wholly unconnected with the State, that draw the collectivity in this or that direction. Frequently it is a case of the State coming under their pressure, rather than itself giving the impulse to them. In this way a whole psychic life is diffused throughout the society. But it is a different one that has a fixed existence in the organ of government. It is here that this other psychic life develops and when in time it begins to have its effect on the rest of the society, it is only in a minor way and by repercussions. When a bill is carried in Parliament, when the government takes a decision within the limits of its competence, both actions, it is true, depend on the general state of social opinion, and on the society. Parliament and the government are in touch with the masses of the nation and the various im·- pressions released by this contact have their effect in deciding them to take this course rather than that. But even if there be this one factor in their decision lying outside themselves, it is none the less true that it is they (Parliament and government) who make this decision and above all it expresses the particular *milieu* where it has its origin. It often happens, too, that there may even be discord between this *milieu* and the nation as a whole, and that decisions taken by the government or parliamentary vote may be valid for the whole community and yet do not square with the state of social opinion. So we may say that there is a collective psychic life, but this life is not diffused throughout the entire social body: although collective, it is localised in a specific organ. And this localisation does not come about simply through concentration on a given point of a life having its origins outside this point. It is in part at this very point that it has its beginning. When the State takes thought and makes a decision, we must not say that it is the society that thinks and decides through the State, but that the State thinks and decides for it. It is not simply an instrument for canalizing and concentrating. It is, in a certain sense, the organizing centre of the secondary groups themselves.

Let us see how the State can be defined. It is a group of

officials *sui generis*, within which representations and acts of volition involving the collectivity are worked out, although they are not the product of collectivity. It is not accurate to say that the State embodies the collective consciousness, for that goes beyond the State at every point. In the main, that consciousness is diffused: there is at all times a vast number of social sentiments and social states of mind (*états*) of all kinds, of which the State hears only a faint echo. The State is the centre only of a particular kind of consciousness, of one that is limited but higher, clearer and with a more vivid sense of itself. There is nothing so obscure and so indefinite as these collective representations that are spread throughout all societies—myths, religious or moral legends, and so on. . . . We do not know whence they come nor whither they are tending; we have never had them under examination. The representations that derive from the State are always more conscious of themselves, of their causes and their aims. These have been concerted in a way that is less obscured. The collective agency which plans them realizes better what it is about. There too, it is true, there is often a good deal of obscurity. The State, like the individual, is often mistaken as to the motives underlying its decisions, but whether its decisions be ill motivated or not, the main thing is that they should be motivated to some extent. There is always or at least usually a semblance of deliberation, an understanding of the circumstances as a whole that make the decision necessary, and it is precisely this inner organ of the State that is called upon to conduct these debates. Hence, we have these councils, these regulations, these assemblies, these debates that make it impossible for these kinds of representation to evolve except at a slow pace. To sum up, we can therefore say that the State is a special organ whose responsibility it is to work out certain representations which hold good for the collectivity. These representations are distinguished from the other collective representations by their higher degree of consciousness and reflection.

We may perhaps feel some surprise at finding excluded from this definition all idea of action or execution or achievement of plans outside the State. Is it not generally held that this part of the State (at all events the part more precisely called the government), has the executive power? This view, however, is

altogether out of place: the State does not execute anything.
The Council of ministers or the sovereign do not themselves
take action any more than Parliament: they give the orders for
action to be taken. They co-ordinate ideas and sentiments,
from these they frame decisions and transmit these decisions
to other agencies that carry them out: but that is the limit of
their office. In this respect there is no difference between
Parliament (or the deliberative assemblies of all kinds surrounding the sovereign or head of State) and the government in the
exact meaning of the term, the power known as executive.
This power is called executive because it is closest to the
executive agencies, but it is not to be identified with them.
The whole life of the State, in its true meaning, consists not
in exterior action, in making changes, but in deliberation,
that is, in representations. It is others, the administrative bodies
of all kinds, who are in charge of carrying out the changes.
The difference between them and the State is clear: this
difference is parallel to that between the muscular system
and the central nervous system. Strictly speaking, the State is
the very organ of social thought. As things are, this thought is
directed towards an aim that is practical, not speculative. The
State, as a rule at least, does not think for the sake of thought
or to build up doctrinal systems, but to guide collective conduct. None the less, its principal function is to think.

But what is the direction of this thought? or, in other words,
what end does the State normally pursue and therefore should
it pursue, in the social conditions of the present day? This is the
question that still remains, and only when it has been solved
can we understand what the citizen's duty is to the State and
the State's to the citizen. Two conflicting solutions are usually
given to this problem.

First, there is that known as individualistic, as expounded
and defended by Spencer and the classical economists on the
one hand and by Kant, Rousseau and the spiritualistic school
on the other. The purpose of society, it is held, is the individual
and for the sole reason that he is all that there is that is real in
society. Since it is only an aggregate of individuals, it can have
no other aim than the development of individuals. Indeed, by
the very fact of the association, society makes human activity
more productive in the realm of science, the arts and industry.

Thanks to this greater yield, the individual finds more abundant nourishment, material and moral as well as for the intellect and so he thrives and develops. But the State is not of itself a producer. It adds nothing and can add nothing to this wealth of all kinds that the society stores up and that the individual benefits from. What then is the part it should play? The answer is, to ward off certain ill effects of the association. The individual in himself has from birth certain rights, by the sole fact that he exists. He is, says Spencer, a living being, therefore he has the right to live, the right not to be obstructed by any other individual in the regular functioning of his organism. He is, says Kant, a moral personality, by virtue of which he is endowed with a particular character that calls for respect, whether in his civil status or in that status known as natural. These inborn rights, in whatever way one may understand or explain them, are in some respects shaped by the association. Any person, in his dealings with me, by the very fact that we are in social intercourse, may either threaten my existence or obstruct the regular activity of my vital forces, or, to use the language of Kant, he may be lacking in the respect due to me or transgress in me the rights of the moral individual that I am. Therefore some agency must be assigned to the precise task of watching over the maintenance of these individual rights. For if the society can and should add something to what I hold by natural endowment (and held before ever society had any hand in founding such rights in my behalf), it must first of all prevent their being impinged upon: otherwise it has no further *raison d'être*. That is a minimum, to which the society need not confine itself, but below which it must not allow one to fall, even if it were to offer us some luxury in place of it, which could have no value if the necessity were lacking in whole or in part. Likewise, many thinkers, of divergent schools, have held that the prerogatives of the State should be limited to administering a wholly negative justice. Its role was to be reduced more and more to preventing unlawful trespass of one individual on another and to maintain intact in behalf of each one the sphere to which he has a right solely because he is what he is. It is true they know well enough that in fact the functions of the State in the past were far more numerous. But they attribute this number of prerogatives to those conditions in which societies exist that have not reached a sufficiently high stage of civilization. In these the state of war is sometimes chronic, and always recurring. War, of course, leads to a disregard of individual rights. It demands severe discipline and this discipline in turn presupposes a strongly entrenched authority. It is from this source there comes the sovereign power over individuals that is so often lodged in the State. The State, on the strength of this authority, has intervened in fields which by their nature should remain alien to it. It controls religious beliefs, industry and so on by regulation. But this unwarranted spread of its influence can only be justified wherever war plays an important part in the life of a people. The more it retreats, the less often it occurs, the more possible and imperative it becomes to disarm the State. War has not yet entirely gone out and there are still threats of international rivalry: so the State, even to-day, still has to preserve a measure of its former prerogatives. But here, in war, we have only something of an anomalous survival, and gradually the last traces of it are bound to be wiped out.

At the point we have reached, there is no need to refute this theory in detail. First, obviously, it does not agree with the facts. As we read on in history, we see the functions of the State multiplying as they increase in importance. This development of the functions is made materially perceptible by the parallel development of the organ itself. What a far cry from the instrument of government in a society such as our own to what it was in Rome or in a Red Indian tribe. In the one, a score of ministries with all their interlocking, side by side with huge assemblies whose very structure is infinitely complex, and over all, the head of State with his own particular administrative departments. In the other, a prince or a few magistrates, some counsellors aided by secretaries. The social brain, like the human brain, has grown in the course of evolution. And yet war during this time, except for some passing setbacks, has become more and more intermittent and less common. We should therefore consider as radically abnormal this theory of a progressive development of the State and the unbroken expansion of its functions, say, in the administration of justice; and given the continuity and regular course of this expansion throughout history, such a hypothesis is untenable. We should need supreme

53

confidence in the force of our own dialectic to condemn as unhealthy such constant and general changes in the name of a particular system. There is not one State whose budget is not visibly becoming inflated. The economists see in this the deplorable result of a clear case of faulty reasoning and they moan over the prevailing blindness. It would perhaps be a better idea to consider a tendency so universally inevitable as regular and normal: always excepting of course certain passing excesses and abuses, which no one would deny.

Apart from this doctrine, it remains to say that the State has other aims and offices to fulfil than watching over individual rights. But here we are likely to be faced by a solution quite contrary to the one we have just been examining—one I might perhaps call the mystic solution. It is this one that we find more systematically set out in the social theories of Hegel than elsewhere, at any rate in some respects. Seen from this point, it is argued that every society has an aim superior to individual aims and unrelated to them. It is held that the part of the State is to pursue the carrying out of this truly social aim, whilst the individual should be an instrument for putting into effect the plans he has not made and that do not concern him. It is to the glory of the society, for its greatness and for its riches he has to labour: he has to find recompense for his pains in the sole fact that as a member of the society he has some sort of share in the benefits he has helped to win. He does receive some of the rays of this glory; a reflection of this splendour does spread to him and that is enough to hold his interest in the aims that lie beyond his reach. This argument deserves to hold our attention all the more because its interest is not solely speculative or historic; the existing confusion in ideas gives it strength and it is about to enjoy a kind of revival. Our own country, which has hitherto been deaf to this argument, now seems ready to welcome it. Since the old individual aims I have just set forth no longer suffice, there are those who throw themselves in despair back on the opposite faith and, renouncing the cult of the individual which was enough for our fathers, they try to revive the cult of the City State in a new guise.

V

CIVIC MORALS (Continued)

THERE is no doubt, in the case of very many societies, what was the true nature of the aims pursued by the State. To keep on expanding its power and to add lustre to its fame—this was the sole or main object of public activity. Individual interests and needs did not come into the reckoning. The ingrained religious character of the political system of societies makes us appreciate this indifference of the State for what concerns the individual. The destiny of a State was closely bound up with the fate of the gods worshipped at its altars. If a State suffered reverses, then the prestige of its gods declined in the same measure—and vice versa. Public religion and civic morals were fused: they were but different aspects of the same reality. To bring glory to the City was the same as enhancing the glory of the gods of the City: it worked both ways. Now, the phenomena in the religious sphere can be recognized because they are wholly unlike those of the human order. They belong to a world apart. The individual *qua* individual is part of the profane world, whilst the gods are the very nucleus of the religious world, and between these two worlds there is a gulf. The gods are, in their substance, different from men: they have other ideas, other needs and an existence with no likeness to that of men. Anyone who holds that the aims of the political system were religious and the religious aims political, might as well say that there was a cleavage between the aims of the State and the ends pursued by individuals on their own. How then came it that the individual could thus occupy himself with the pursuit of aims which were to such a degree foreign to his own private concerns? The answer is this: his private concerns were relatively unimportant to him and his personality and everything that hung on it had but slight moral weight. His personal views, his private beliefs and all his diverse aspirations as an individual seemed insignificant factors. What was prized by all, were the beliefs held in common, the collective aspirations, the popular traditions and the symbols that were an expression of them. That being so, it was gladly and without any demur that the individual yielded to the instrument by which the aims of no immediate concern to himself were secured. Absorbed, as he was, into the mass of society, he meekly gave way to its pressures and subordinated his own lot to the destinies of collective existence without any sense of sacrifice. This is because his particular fate had in his own eyes nothing of the meaning and high significance that we nowadays attribute to it. If we are right in that estimate, it was in the nature of things that it should be so; societies could only exist at that time by virtue of this subservience.

But the further one travels in history, the more one is aware of the process of change. In the early stage, the individual personality is lost in the depths of the social mass and then later, by its own effort, breaks away. From being limited and of small regard, the scope of the individual life expands and becomes the exalted object of moral respect. The individual comes to acquire ever wider rights over his own person and over the possessions to which he has title; he also comes to form ideas about the world that seem to him most fitting and to develop his essential qualities without hindrance. War fetters his activity, diminishes his stature and so becomes the supreme evil. Because it inflicts undeserved suffering on him, he sees in it more and more the supreme form of moral offence. In the light of this, it is utterly inconsistent to require from him the same subordination as before. One cannot make of him a god, a god above all others, and at the same time an instrument in the hands of the gods. One cannot make of him the paramount end and reduce him to the role of means. If he be the moral reality, then it is he who must serve as the pole-star for public as well as private conduct. It should be the part of the State to try to bring his innate qualities to the light. Shall we find some people saying that the cult of the individual is a superstition of which we ought to rid ourselves? That would be to go against all the lessons of history: for as we read on, we find the human person tending to gain in dignity. There is no rule more soundly established. For any attempt to base social institutions on the opposite principle is not feasible and could be convincing only for a moment: we cannot force things to be other than they are. We cannot undo the individual having become what he is—an autonomous centre of activity, an impressive system of personal forces whose energy can no more be destroyed than that of the cosmic forces. It would be just as impossible to transform our physical atmosphere, in the midst of which we breathe and have our being.

Do we not arrive here at a contradiction that cannot be resolved? On the one hand we establish that the State goes on developing more and more: on the other, that the rights of the individual, held to be actively opposed to those of the State, have a parallel development. The government organ takes on an ever greater scale, because its function goes on growing in importance and because its aims, that are in line with its own activity, increase in number; yet we deny that it can pursue aims other than those that connern the individual. Now, these aims are by definition held to belong to individual activity. If, as we suppose, the rights of the individual are inherent, the State does not have to intervene to establish them, that is, they do not depend on the State. But then, if they do not, and are outside its competence, how can the cadre of this competence go on expanding, in face of the fact that it must less and less take in things alien to the individual?

The only way of getting over the difficulty is to dispute the postulate that the rights of the individual are inherent, and to admit that the institution of these rights is in fact precisely the task of the State. Then, certainly, all can be explained. We can understand that the functions of the State may expand, without any diminishing of the individual. We can see too that the individual may develop without causing any decline of the State, since he would be in some respects the product himself of the State, and since the activity of the State would in its nature be liberating to him. Now, what emerges, on the evidence of the facts, is that history gives sound authority for this relation of cause and effect as between the progress of moral individualism and the advance of the State. Except for the abnormal cases we shall discuss later, the stronger the State, the more the individual is respected. We know that the Athenian State was far less tightly constructed than Rome, and it is clear that the Roman State, again, more especially the City State, was built on very simple lines, compared with the great centralized States of our own day. Progress in concentration of government in the Roman City took a different course from that in any of the Greek Cities, and the unit of the State had a different emphasis. This point we settled last year. One outstanding fact makes us aware of this difference: in Rome, the direction of religious practices was in the hands of the State. In Athens, it was dispersed amongst the many sacerdotal colleges. Nothing is to be found in Athens corresponding to the Roman Consul, in whose hands all governmental power was concentrated. The administration in Athens was distributed amongst an unco-ordinated crowd of various officials. Each of the group elements that made up the society—clans, *phratries* and tribes—had preserved an autonomy far greater than in Rome, where they were very soon absorbed in the social mass. In this respect, the distance that stretches between the modern European States and the Greek or Italian States is obvious. Now, individualism had a different development in Rome as compared with Athens. In Rome, the lively sense of the respect due to the person was expressed, first, in recognized terms affirming the dignity of the Roman citizen and, secondly, in the liberties which were its distinguishing juridical features.

This is one of the points on which Ihering has helped to throw a sharp light. We are in the same case in respect of freedom of thought. But remarkable as Roman individualism may be, it is slight enough compared to that which developed within Christian societies. The Christian form of religion is an inward one: it consists of inward faith rather than outward observances, for a deeply held faith eludes any external constraint. In Athens, intellectual development—scientific and philosophical—was far greater than in Rome. Now, it is held that science and philosophy and collective thinking develop in the same way as individualism. True, they very often accompany it, but that is not inevitably so. In India, Brahmanism and Buddhism have a very learned and very subtle metaphysic: the Buddhist religion rests on a whole theory of the world. The sciences were developed to a high degree in the temples of Egypt. We know, however, that in the case of both India and Egypt, there was an almost complete absence of individualism. It is this fact more than any other that goes to prove the pantheistic nature of these metaphysics and religions: they attempted to give the pantheism a kind of rational and charted formula. Clearly, a pantheistic faith is not possible where individuals have a lively sense of their individuality.

Again, letters and philosophy were widely pursued in the medieval monasteries. That was because intensity of speculation, in the individual as in the society, is in fact in inverse ratio to practical activity. When we find activity in the practical field falling below the normal in any one section of society, for some reason or other, then the intellectual forces will develop all the more and flow into the space thus left open to them. So it was with the priests and monks, especially in the contemplative religions. From another angle, we know too that for the Athenian, the matter of practical life was reduced to something insignificant. He lived a life of leisured pursuits. In such a setting there comes a remarkable flowering of science and philosophy. Once they flower, they may, to be sure, inspire an individualist movement, but we cannot say they derive from it. It is possible, of course, that speculation, opening out in this way, may not have this result and that it remains in its essence conservative. In that case it is taken up with making a theory of the state of things as they exist or perhaps with a commentary on it. Such, in the main, is the nature of sacerdotal speculation: and even Greek speculation as a whole had this same tendency over a long period. The political and moral theories of Plato and Aristotle hardly do more than reflect in their systems the political structure of Sparta and Athens respectively.

Finally, one last reason that prevents our measuring the degree of individualism in a country by the development reached in the faculties of speculative thought. This is, that individualism is not a theory: it lies in the region of practice, not in that of speculation. For it to be true individualism, it must make its mark on morals and social institutions. There are times, too, when it dissipates itself entirely, as it were, in speculative dreaming instead of getting through to reality and initiating that whole collection of customs and institutions that would be adequate to its needs. It is then we see systems come into view that reveal social ideals looking to a more highly developed individualism. That, however, remains a mere desideratum, since the conditions needed to make it a reality are lacking. Is this not rather the case with our own French individualism? It was expressed theoretically in the Declaration of the Rights of Man, although in exaggerated form; it is, however, far from having any deep roots in the country. The proof of this is seen in the extreme ease with which we have accepted an authoritarian regime several times in the course of this century—regimes which in reality rest on principles that are a long way from individualism. The old habits persist more than we think, more than we should like, in spite of the letter of our moral code. The reason is, that in order to set up an individualistic moral code, it is not enough to assert it or to translate it into fine systems. Society, rather, must be so ordered that this set-up is made feasible and durable. Otherwise, it remains in a vague doctrinaire state.

History seems indeed to prove that the State was not created to prevent the individual from being disturbed in the exercise of his natural rights: no, this was not its role alone—rather, it is the State that creates and organizes and makes a reality of these rights. And indeed, man is man only because he lives in society. Take away from man all that has a social origin and nothing is left but an animal on a par with other animals. It is society that has raised him to this level above physical nature: it has achieved this result because association, by grouping the individual psychic forces, intensifies them. It carried them to a degree of energy and productive capacity immeasurably greater than any they could achieve if they remained isolated one from the other. Thus, a psychic life of a new kind breaks away which is richer by far and more varied than one played out in the single individual alone. Further, the life thus freed pervades the individual who shares in it and so transforms him. Whilst society thus feeds and enriches the individual nature, it tends, on the other hand, at the same time inevitably to subject that nature to itself and for the same reason. It is precisely because the group is a moral force greater to this extent than that of its parts, that it tends of necessity to subordinate these to itself. The parts are unable *not* to fall under its domination. Here there is a law of moral mechanics at work, which is just as inevitable as the laws of physical mechanics. Any group which exercises authority over its members by coercion strives to model them after its own pattern, to impose on them its ways of thinking and acting and to prevent any dissent.

Every society is despotic, at least if nothing from without supervenes to restrain its despotism. Still, I would not say that there is anything artificial in this despotism: it is natural because it is necessary, and also because, in certain conditions, societies cannot endure without it. Nor do I mean that there is anything intolerable about it: on the contrary, the individual does not feel it any more than we feel the atmosphere that weighs on our shoulders. From the moment the individual has been raised in this way by the collectivity, he will naturally desire what it desires and accept without difficulty the state of subjection to which he finds himself reduced. If he is to be conscious of this and to resist it, individualist aspirations must find an outlet, and that they cannot do in these conditions.

But for it to be otherwise, we may say, would it not be enough for the society to be on a fairly large scale? There is no doubt that when it is small—when it surrounds every individual on all sides and at every moment—it does not allow of his evolving in freedom. If it be always present and always in action, it leaves no room to his initiative. But it is no longer in the same case when it has reached wide enough dimensions. When it is made up of a vast number of individuals, a society can exercise over each a supervision only as close and as vigilant and effective as when the surveillance is concentrated on a small number. A man is far more free in the midst of a throng than in a small coterie. Hence it follows that individual diversities can then more easily have play, that collective tyranny declines and that individualism establishes itself in fact, and that, with time, the fact becomes a right. Things can, however, only have this course on one condition: that is, that inside this society, there must be no forming of any secondary groups that enjoy enough autonomy to allow of each becoming in a way a small society within the greater. For then, each of these would behave towards its members as if it stood alone and everything would go on as if the full-scale society did not exist. Each group, tightly enclosing the individuals of which it was made up, would hinder their development; the collective mind would impose itself on conditions applying to the individual. A society made up of

adjoining clans or of towns or villages independent in greater
or lesser degree, or of a number of professional groups, each
one autonomous in relation to the others, would have the effect
of being almost as repressive of any individuality as if it were
made up of a single clan or town or association. The formation
of secondary groups of this kind is bound to occur, for in a great
society there are always particular local or professional interests
which tend naturally to bring together those people with whom
they are concerned. There we have the very stuff of associations of a special kind, of guilds, of coteries of every variety;
and if there is nothing to offset or neutralize their activity, each
of them will tend to swallow up its members. In any case, just to
take the domestic society: we know its capacity to assimilate
when left to itself. We see how it keeps within its orbit all those
who go to make it up and are under its immediate domination.
(At any rate, if secondary groups of this sort are not formed,
at least a collective force will establish itself at the head of the
society to govern it. And if this collective force itself stands
alone, if it has only individuals to deal with, the same law of
mechanics will make those individuals fall under its domination).

In order to prevent this happening, and to provide a certain
range for individual development, it is not enough for a society
to be on a big scale; the individual must be able to move with
some degree of freedom over a wide field of action. He must not
be curbed and monopolised by the secondary groups, and these
groups must not be able to get a mastery over their members
and mould them at will. There must therefore exist above these
local, domestic—in a word, secondary—authorities, some overall authority which makes the law for them all: it must remind
each of them that it is but a part and not the whole and that it
should not keep for itself what rightly belongs to the whole.
The only means of averting this collective particularism and all
it involves for the individual, is to have a special agency with the
duty of representing the overall collectivity, its rights and its
interests, vis-à-vis these individual collectivities.

These rights and these interests merge with those of the
individual. Let us see why and how the main function of the
State is to liberate the individual personalities. It is solely
because, in holding its constituent societies in check, it prevents

them from exerting the repressive influences over the individual
that they would otherwise exert. So there is nothing inherently
tyrannical about State intervention in the different fields of
collective life; on the contrarary, it has the object and the effect
of alleviating tyrannies that do exist. It will be argued, might
not the State in turn become despotic? Undoubtedly, provided
there were nothing to counter that trend. In that case, as the
sole existing collective force, it produces the effects that any
collective force not neutralized by any counter-force of the
same kind would have on individuals. The State itself then
becomes a leveller and repressive. And its repressiveness
becomes even harder to endure than that of small groups,
because it is more artificial. The State, in our large-scale
societies, is so removed from individual interests that it cannot
take into account the special or local and other conditions in
which they exist. Therefore when it does attempt to regulate
them, it succeeds only at the cost of doing violence to them and
distorting them. It is, too, not sufficiently in touch with individuals in the mass to be able to mould them inwardly, so
that they readily accept its pressure on them. The individual
eludes the State to some extent—the State can only be effective
in the context of a large-scale society—and individual diversity
may not come to light. Hence, all kinds of resistance and distressing conflicts arise. The small groups do not have this
drawback. They are close enough to the things that provide
their *raison d'être* to be able to adapt their actions exactly and
they surround the individuals closely enough to shape them in
their own image. The inference to be drawn from this comment,
however, is simply that if that collective force, the State, is to
be the liberator of the individual, it has itself need of some
counter-balance; it must be restrained by other collective
forces, that is, by those secondary groups we shall discuss later
on. . . . It is not a good thing for the groups to stand alone,
nevertheless they have to exist. And it is out of this conflict of
social forces that individual liberties are born. Here again we
see the significance of these groups. Their usefulness is not
merely to regulate and govern the interests they are meant to
serve. They have a wider purpose; they form one of the conditions essential to the emancipation of the individual.

It remains a fact that the State is not of its own volition

antagonistic to the individual. It is only through the State that individualism is possible, although it cannot be the means of making it a reality, except in certain precise conditions. We might say that in the State we have the prime mover. It is the State that has rescued the child from patriarchal domination and from family tyranny; it is the State that has freed the citizen from feudal groups and later from communal groups; it is the State that has liberated the craftsman and his master from guild tyranny. It may take too violent a course, but the action becomes vitiated only when it is merely destructive. And that is what justifies the increasing scope of its functions. This concept of the State is, then, an individualistic one, but it does not limit the State to the administration of an entirely prohibitive justice. And in this concept there is recognition of the right and duty of the State to play the widest possible part in all that touches collective life, without however having a *mystique*.[1] For the purpose assigned to the State in this concept is comprehensible to individuals, just as they understand the links between the State and themselves. They may co-operate in this, fully realizing what they are about and the ultimate aim of their actions, because it is a matter that concerns themselves. They may even find themselves in opposition to that aim and thus even become instruments of the State, for it is towards making them a reality that the action of the State tends. And yet they are not (as held by the individualistic utilitarians or the school of Kant) wholes that are self-sufficing and that the State should merely respect, since it is through the State, and the State alone, that they have a moral existence.

[1] N.B. 'without becoming, as it were, a mystic concept of State.'

VI

CIVIC MORALS (Continued)

THE STATE AND THE INDIVIDUAL—PATRIOTISM

WE should now set forth how the State, without pursuing a mystic aim of any kind, goes on expanding its functions. If indeed we work on the premise that the rights of the individual are not *ipso facto* his at birth; that they are not inscribed in the nature of things with such certainty as warrants the State in endorsing them and promulgating them; that, on the contrary, the rights have to be won from the opposing forces that deny them; that the State alone is qualified to play this part—then it cannot keep to the functions of supreme arbiter and of administrator of an entirely prohibitive justice, as the utilitarian or Kantian individualism would have it. No, the State must deploy energies equal to those for which it has to provide a counter-balance. It must even permeate all those secondary groups of family, trade and professional association, Church, regional areas and so on . . . which tend, as we have seen, to absorb the personality of their members. It must do this, in order to prevent this absorption and free these individuals, and so as to remind these partial societies that they are not alone and that there is a right that stands above their own rights. The State must therefore enter into their lives, it must supervise and keep a check on the way they operate and to do this it must spread its roots in all directions. For this task, it cannot just withdraw into the tribunals, it must be present in all spheres of social life and make itself felt. Wherever these particular collective forces exist, there the power of the State must be, to neutralize them: for if they were left alone and to their own devices, they would draw the individual within their exclusive domination. Now, societies are becoming ever greater in scale and ever more complex: they are made up of circles of increasing diversity, and of manifold

65

agencies, and these already possess in themselves a value to be reckoned. Therefore if it is to fulfil its function, the State, too, must branch out and evolve to the same degree.

It would be easier to understand the need for this whole trend of expansion if we could form a better idea of the elements of these individual rights that the State secures by stages, overcoming the resistances of collective particularism. We may hold (with Spencer and Kant, to quote only the leaders of the school) that these rights derive from the very nature of the individual and only express the conditions necessary to him if he is to be himself. Then we are bound to conceive these rights as defined and determined once and for all, as well as that individual nature which they express and from which they derive. Take any human being: he is endowed with a certain mental and moral constitution; his rights are dependent on that constitution, being implicitly written into it, as it were. We could draw up an exhaustive and final list of them, with omissions no doubt, but there would be nothing indefinite about the list as it stood, and with adequate method, it could give a complete picture. If individual rights are to ensure a free functioning of the life of the individual, it only remains to settle what that life involves, to deduce the rights that must be conceded to the individual. For instance, according to Spencer, life in man presupposes a constant equilibrium between the vital energies and the exterior energies; this means that the process of repair must balance the expenditure of the energy or the wear and tear. Each one of us should therefore receive in exchange for his work a remuneration allowing him to repair the energies consumed by the work. That would be met if contracts were freely made and abided by, for the individual should never yield up what he has made or done in exchange for something of less value. Man, says Kant, is a moral being. His right derives from the moral nature he is endowed with and is thus determined by that very fact. This moral nature makes him inviolable; anything that assails his inviolability is a violation of this right. That is how those who uphold what is called natural right (or the theory of individual right deriving from individual nature) come to represent it as being something universal; that is, as a code that can be laid down once and for all, valid for every period as for every country. And this negative character they try to give to this right makes it, apparently, more easily definable.

But the postulate on which this theory rests has an artificial over-simplification. What lies at the base of individual right is not the notion of the individual as he is, but the way in which society puts the right into practice, looks upon it and appraises it. What matters, is not what the individual is, but how much he counts and on the other hand, what he ought to be. The reason why he has more or fewer rights, certain rights and not others, is not that he is constituted in a particular way; it is because society attributes this or that importance to him and attaches a higher or a lower value to what concerns him. If all that affects the individual affects the society, the society will react against all that might diminish him. This would not only forbid the slightest offences against him, but even more, the society would hold itself bound to work towards increasing his stature and towards his development. If, on the other hand, the individual is held in only moderate regard, the society will be indifferent even to serious outrages on him and will tolerate them. According to ideas current at the time, grave offences will appear as venial or, on the contrary, it may be held that liberal, unfettered expression should not be too much encouraged. Those who believe in that theory of natural right think they can make a final distinction between what is and what is not a right. However, a closer study will show that in reality the dividing line they think they can draw is certainly not definite and depends entirely on the state of public opinion. Spencer remarks that the remuneration shall be equal to the value of the labour—that this must be and suffices. But how is this balance to be settled? This value is a matter of opinion. It is said that the contracting parties must decide this, provided that they decide freely. But again, what does this freedom consist of? Nothing has fluctuated so much in the course of time as the idea of freedom of contract. With the Romans, the contract came into force at the moment its text was declared and it was the phrasing of the text that governed the engagements entered into and not the intention behind the words. Later, the intention began to come into the reckoning and the contract made under material duress was no longer held to be regular. Some forms of moral pressure likewise began to be ruled out.

What brought about this development? The answer is, that people began to have a far loftier idea of the human person and the smallest attempt on his freedom became more intelerable. Everything points to this development not having ended yet, and to our becoming even more severe in this matter. Kant declares that the human person should be autonomous. But an absolute autonomy is out of the question. The human person forms part of the physical and social *milieu*; he is bound up with it and his autonomy can be only relative. And then, what degree of autonomy is appropriate to him? It is obvious that the answer depends on the state of mind of the societies— that is, on the state of public opinion. There was a time when material servitude, imposed in certain conditions, seemed in no wise immoral; we have abolished it, but how many forms of moral servitude still survive? Can we say that a man who has nothing to live on governs himself, that he is master of his actions? Which kinds of subordination, then, are legitimate and which unlawful? There is no final answer to these problems.

The rights of the individual, then, are in a state of evolution: progress is always going on and it is not possible to set any bounds to its course. What yesterday seemed but a kind of luxury becomes overnight a right precisely defined. The task incumbent on the State, then, has no limits. It is not merely that it has to work out a definite ideal that sooner or later has to be attained, and that finally. But the field open to its moral activity is immeasurable. There is no reason why it should ever cease to be so or the work ever be considered as finished. Everything indicates that we are becoming more alive to what touches on the human personality. Even if we fail to foresee the coming changes along these lines and in this spirit, our lack of imagination does not warrant our shutting our eyes to them. Besides, there are already many changes that we can foresee will be necessary. These considerations explain more clearly the continuous advance of the State and its justification, to some extent: they allow us to assume that far from being some kind of passing anomaly, this advance is bound to go on indefinitely in the future.

Meanwhile, it is now easier to see there was no exaggeration in saying that our moral individuality, far from being antagonistic to the State, has on the contrary been a product of it.

It is the State that sets it free. And this gradual liberation does not simply serve to fend off the opposing forces that tend to absorb the individual: it also serves to provide the *milieu* in which the individual moves, so that he may develop his faculties in freedom. There is nothing negative in the part played by the State. Its tendency is to ensure the most complete individuation that the state of society will allow of. Far from its tyrannizing over the individual, it is the State that redeems the individual from the society. But whilst this aim is essentially positive, it has nothing transcendental about it for the individual consciousness, for it is an aim that is also essentially human. There is no difficulty in understanding its appeal, for ultimately it concerns ourselves. Individuals can become instruments of the State without any inconsistency, since the action of the State is towards giving them reality. We do not, even so, follow Kant and Spencer in making them into absolutes, as it were, almost self-sufficing, or into egotisms knowing only self-interest. For although this aim concerns them all, it cannot in the main be identified with the aim of any one of them in particular. It is not this or that individual the State seeks to develop, it is the individual *in genere*, who is not to be confused with any single one of us. And whilst we give the State our co-operation—and it could do nothing without it—we do not become the agents of a purpose alien to us; we do not give up the pursuit of an impersonal aim which belongs to a region above all our own private aims but which nevertheless has close ties with them. On the one hand, our concept of the State has nothing mystic about it, and yet, it is still in its essence individualistic.

The fundamental duty of the State is laid down in this very fact : it is to persevere in calling the individual to a moral way of life. I say fundamental duty, for civic morals can have no pole-star for guide except moral causes. If the cult of the human person is to be the only one destined to survive, as it seems, it must be observed by the State as by the individual equally. This cult, moreover, has all that is required to take the place of the religious cults of former times. It serves quite as well as they to bring about that communion of minds and wills which is a first condition of any social life. It is just as simple for men to draw together to work for the greatness of man as it is to work to the glory of Zeus or Jehovah or Athena. The whole difference of this religion, as it affects the individual, is that the god of its devotion is closer to his worshippers. But although not far removed, he does nevertheless still transcend them, and the rôle of the State in this respect is what it was formerly. It rests with it, shall we say, to organize the cult, to be the head of it and to ensure its regular working and development.

Shall we say that this is the sole duty incumbent on the State and that its whole activity should be directed into this channel? It might be so if every society lived in isolation, without having to fear any hostile acts. But we know that international competition has not yet ended and that even 'civilised' States still live to some extent on a war footing in their inter-relations. They threaten one another, and since the first duty of the State towards its members is to preserve intact the collective entity they make up, it must to that extent organize itself accordingly. It must be ready to defend itself, perhaps even to attack if it feels menaced. This whole form of organization presupposes a different kind of moral discipline from that implicit in the cult of the human being. It has an entirely different cast of direction. Its goal is a national collectivity and not the individual. It is a survival of the discipline of other days, since the former conditions of existence have not yet ceased to operate. There are, then, two diverging currents flowing through our moral life. It would be failing to recognize the existing state of affairs, if we wished to reduce this duality to unity here and now, if we wished to do away with all these institutions, all these practices inherited from the past, straight away, whilst the conditions that created them still survive. Just as we cannot make it a fact that individual personality shall not have reached the stage of evolution that it has, so we cannot make it a fact that international competition shall not have preserved a military form. Hence come these duties of an entirely different nature for the State. Nothing even warrants our assuming that some part of them will not always continue to exist. As a rule, the past never disappears entirely. Something of it always survives into the future. That said, it remains to add that, as we progress, so these duties (as explained)— once fundamental and essential, become secondary and anomalous: that is, always providing that nothing unusual occurs and there are no fortuitous setbacks. Once, the action of the State was directed entirely outwards: now, inevitably, it tends more and more to turn inwards. For it is through this whole structure of the State and through it alone, that society can succeed in achieving the aim it has to put foremost. And there is not likely to be any lack of substance to work on here. The planning of the social *milieu* so that the individual may realise himself more fully, and the management of the collective apparatus in a way that will bear less hard on the individual; an assured and amicable exchange of goods and services and the co-operation of all men of good will towards an ideal they share without any conflict: in these, surely, we have enough to keep public activity fully employed. No European country is free of internal problems and difficulties, and as we go on, so will these problems multiply. That is so, because, as social life becomes more complex, so does the working of its functions become more delicate. Further, since the more highly developed systems are precariously balanced and need greater care if they are to be kept going, societies will have a growing need to concentrate their energies on themselves to husband their strength, instead of expending them outwards in violent demonstrations.

This is where Spencer's arguments have some plausibility. He saw clearly that the receding of war and of the social forms or methods bound up with it was certain to affect the life of all societies very deeply. But it does not follow that this recession leaves no other sustenance for social life than economic interests and that there must inevitably be a choice between militarism and commercialism. If, to use his expression, the organs of depredation tend to disappear, this does not mean that the organs of a vegetative system should entirely take their place, nor that the social organs should one day be reduced to no more than a vast digestive apparatus. There is an inward activity that is neither economic nor commercial and this is moral activity. Those forces that turn from the outward to the inward are not simply used to produce as much as possible and to add to creature comfort, but to organize and raise the moral level of society, to uphold this moral structure and to see that it goes on developing. It is not merely a matter of increasing the exchanges of goods and services, but of seeing that they are done by rules that are more just; it is not simply that everyone should have access to rich supplies of food and drink. Rather, it is that each one should be treated as he deserves, each be freed from an unjust and humiliating tutelage, and that, in holding to his fellows and his group, a man should not sacrifice his individuality. And the agency on which this special responsibility lies is the State. So the State does not inevitably become either simply a spectator of social life (as the economists would have it), in which it intervenes only in a negative way, or (as the socialists would have it), simply a cog in the economic machine. It is above all, supremely the organ of moral discipline. It plays this part at the present time as it did formerly, although the discipline has changed. (Here we see the error of the socialists.)

The conclusion that we reach here gives an indication how one of the gravest conflicts of our day might be solved. By this I mean the conflict that has come about between equally high-minded kinds of sentiment—those we associate with a national ideal and the State that embodies it, and those we associate with the human ideal and mankind in general—in a word, between patriotism and world patriotism. This conflict was unknown to the ancient world, because in those days one cult alone was possible: this was the cult of the State, whose public religion was but the symbolic form of that State. For the worshippers there was therefore nothing to allow of choice or hesitation. They could conceive of nothing above the State, above its fame and greatness. But since then, things have changed. No matter how devoted men may be to their native land, they all to-day are aware that beyond the forces of national life there are others, in a higher region and not so transitory, for they are unrelated to conditions peculiar to any given political group and are not bound up with its fortunes. There is something more universal and more enduring. It is true to say that those aims that are the most general and the most unchanging are also the most sublime. As we advance in evolution, we see the ideals men pursue breaking free of the local or ethnic conditions obtaining in a certain region of the world or a certain human group, and rising above all that is particular and so approaching the universal. We might say that the moral forces come to have a hierarchic order according to their degree of generality or diffusion.

Thus, everything justifies our belief that national aims do not lie at the summit of this hierarchy—it is human aims that are destined to be supreme.

On this basis, it has sometimes been held that patriotism could be regarded simply as a survival that would disappear before long. Here, however, we face another problem. In fact, man is a moral being only because he lives within established societies. There are no morals without discipline and authority, and the sole rational authority is the one that a society is endowed with in relation to its members. Morals do not look like obligations to us, that is, do not seem like morals to us—and therefore we can have no sense of duty—unless there exist about us and above us a power which gives them sanction. Not that the material sanction covers the whole of the duty, but it is the outward sign by which this is recognized, and manifest evidence that there is something above us to which we are subordinate. It is true that the believer is free to make an image of this power for himself in the shape of a superhuman being, inaccessible to reason or science. But for the theme under discussion, we need not debate the hypothesis or examine what is and what is not well founded about the symbol. The fact that shows us to what degree a social structure is necessary to morality is that any disorganization, any tendency to political anarchy, is accompanied by a rise in immorality. This is not solely because the criminal has a better chance of escaping punishment; it is that in general the sense of duty is weakened, because men no longer have a strong sense of there being anything above them to which they are subject. Now, patriotism is precisely the ideas and feelings as a whole which bind the individual to a certain State. If we suppose it to have weakened or to have ceased to exist, where is an individual to find this moral authority, whose curb is to this extent salutary? If there is no clearly defined society there with a consciousness of itself to remind him continually of his duties and to make him realize the need for rules, how should he be aware of all this? Let us take those who believe that morals themselves are inborn and exist *a priori* in the consciousness of each one, and who believe, too, that a man has only to look within his own breast to know what they consist of and needs only a little goodwill to understand that he should submit to them. To these, the State would indeed appear something entirely exterior to morals and therefore it seems that it might lose its dominion without there being any loss to morality. But since we know that morals are a product of the society, that they permeate the individual from without and that in some respects they do violence to his physical nature and his natural temperament, we can understand the better that morals are what the society is and that they have force only so far as the society is organized. At the present day, the State is the highest form of organized society that exists. Some forms of belief in a world State, or world patriotism do themselves get pretty close to an egotistic individualism. Their effect is to disparage the existing moral law, rather than to create others of higher merit. It is for this reason that so many minds resist these tendencies, though realizing that they have something logical and inevitable.

There might indeed be a solution of the problem in theory: this is to imagine humanity in its entirety organized as a society. Need we say that such an idea, whilst not altogether beyond realization, must be set in so distant a future that we can leave it out of our present reckoning. A confederation of European States, for instance, is advanced, but vainly, as a half-way course to achieving societies on a bigger scale than those we know to-day. This greater federation, again, would be like an individual State, having its own identity and its own interests and features. It would not be humanity.

There is however a means of reconciling the two ideas. That is, for the national to merge with the human ideal, for the individual States to become, each in their own way, the agencies by which this general idea is carried into effect. If each State had as its chief aim, not to expand, or to lengthen its borders, but to set its own house in order and to make the widest appeal to its members for a moral life on an ever higher level, then all discrepancy between national and human morals would be excluded. If the State had no other purpose than making men of its citizens, in the widest sense of the term, then civic duties would be only a particular form of the general obligations of humanity. It is this course that evolution takes, as we have already seen. The more societies concentrate their energies inwards, on the interior life, the more they will be diverted from the disputes that bring a clash between cosmopolitism—or world patriotism, and patriotism; as they grow in size and get greater complexity, so will they concentrate more and more on themselves. Here we see how the advent of societies on an even bigger scale than those we know will constitute an advance in the future.

So that what breaks down the paradox is the tendency of patriotism to become, as it were, a fragment of world patriotism. It is a different concept of it that so often leads to conflict. True patriotism, it seems, is only exhibited in forms of collective action directed towards the world without; it seems to us as if we could only show loyalty to our own patriotic or national group at times when it is at strife with some other group. True, these external crises yield plenty of occasions for brilliantly devoted service.

But alongside this patriotism there is another kind, more given to silence but whose effective action is also more sustained; this patriotism is directed towards the interior affairs of the society and not its exterior expansion. It in no wise excludes any national pride: the collective personality and the individual personalities alike can have no existence without an awareness of themselves, of what they are, and this awareness has always something personal. As long as there are States, so there will be national pride, and nothing can be more warranted. But societies can have their pride, not in being the greatest or the wealthiest, but in being the most just, the best organized and in possessing the best moral constitution. To be sure, we have not yet reached the point when this kind of patriotism could prevail without dissent, if indeed such a time could ever come.

VII

CIVIC MORALS (Continued)

FORM OF THE STATE—DEMOCRACY

THE respective duties of State and citizen vary according to the particular form taken by a State. These forms are not the same in what is known as an aristocracy, a democracy or a monarchy. It is therefore of importance to know what these different forms represent and what the origin and basis are of the one that is becoming fairly general in European societies. It is only on these terms that we can understand the origin and basis of our civic duties of the present day.

Ever since the time of Aristotle, States have been classified according to the number of those who have a part in the government. "When the people taken as a whole have sovereign power" says Montesquieu, "it is a democracy. When the sovereign power is in the hands of a section of the people, it is called an aristocracy." (*De l'Esprit des Lois*, Bk II, ch. II.) The monarchic government is one in which a single individual governs. For Montesquieu, however, it is only a true monarchy if the king governs according to fixed and established laws. When, on the other hand, "a single individual, without law or statutes, drives all before him by his will and his caprices," (*Ibid*, Bk. II, ch. I) the monarchy takes the name of despotism. Thus, apart from this matter of there being a constitution or not in existence, it is by the number of those governing that Montesquieu defines the form of a State.

It is true that later on in his book, when he examines the sentiment that is the mainspring of each of these kinds of government, such as honour, valour or fear, he shows that he had a sense of the qualitative differences seen in these varying types of State. For him, however, these qualitative differences are only the result of the purely quantitative differences which we referred to in the first instance and he derives the qualitative from the quantitative. The nature of the sentiment that has to act as driving force for the collective activity is determined by the very number of those governing and so are all the details of organization.

But this way of defining various political forms is as common as it is superficial. To begin with, what are we to understand by 'the number of those governing'? Where does it begin and where end, this governmental organ whose varied forms are to determine the form of the State? Does this mean the aggregate of all those who are appointed to conduct the affairs in general of the country? But all these powers are never or almost never, concentrated in the hands of a single individual. Even though a ruler be absolute, he is always surrounded by advisers and ministers who share these functions for control by rule. Seen from this angle, there are only differences of degree between a monarchy and an aristocracy. A sovereign has always about him a host of officials and dignitaries often as powerful as himself or even more so. Should we consider for our purposes only the highest level of the government organ, the level where the supreme powers are concentrated? I mean those powers which—to use the old term of political theory— appertain to the sovereign. Is it the Head of State alone whom we have in mind? In that case, we should have to keep distinct the State with a single Head, the State with a council of individuals and the State where everyone takes a hand. By this reckoning, seventeenth century France, for example, and a centralized republic like our own present-day France or the United States Republic, would all come under the same heading and would all alike be classed as monarchies. In all these instances there is a single individual at the summit of the monarchy with its officials, and it is his title alone that varies according to the society.

On the other hand, what are we to understand by the words 'to govern'? To govern, it is true, means to exercise a positive control over the course of public affairs. In this respect, a democracy may not be distinguishable from an aristocracy. Indeed, very often, it is the will of the majority that shapes the law, without the views of the minority having the slightest influence. A majority can be as oppressive as a caste. It may even very well happen that the minority is not represented at all in the government councils. Remember, too, that in any case, women, children and adolescents—all those who are prevented from voting for one reason or another—are kept off the electoral lists. The result is that the lists in fact comprise only the minority of the nation. And since those elected represent only the majority in these constituencies, they represent in fact a minority of the minority. In France, out of a population of 38 million, there were in 1893 only 10 million electors; out of these 10 million, 7 million alone made use of their voting rights, and the deputies elected by these 7 million represented only 4,592,000 votes. Taking the whole electorate, 5,930,000 voters were not represented, that is to say, a greater number of voters than those who had returned the deputies elected. Thus, if we confine ourselves to numbers, we have to admit that there has never been a democracy. At the very most we might say— to show where it differs from an aristocracy—that under an aristocratic system the governing minority are established once and for all, whereas in a democracy, the minority that carries the day may be beaten to-morrow and replaced by another. The difference, then, between them is only slight.

Apart from this rather dialectic treatment, there is one historical fact that throws some light on how inadequate these ordinary definitions are.

These definitions would indeed have us approximate types of States that lie, so to speak, at the opposite extremes of evolution. In fact we give the name democracy to those societies where everyone has a share in directing communal life and the word exactly suits the most inferior forms of political society known to us. This description applies to the structure that the English call tribal. A tribe is made up of a certain number of clans. Each clan is ruled by the group itself; when there is a chieftain, his powers are without much force, and the confederation is ruled by a council of representatives. In some respects it is the same system as our own. This resemblance has given weight to the argument that democracy is essentially archaic as a form of society and that an attempt to establish it in present-day societies would be throwing civilization back to its primitive beginnings and reversing the course of history. It is these lines of thought that are sometimes used to draw a parallel between socialist planning and the economic life of communism in the ancient world, in order to demonstrate its alleged futility. We must recognize that in both cases the conclusion would be justified if the postulate were right, that is, if the two forms of social structure here assumed to be the same, were in very fact identical. True, there is no form of government to which the same criticism might not apply, at least, if we confine ourselves to the foregoing definitions. Monarchy is hardly less archaic than democracy. Very often clans or federated tribes were brought together under the hand of an absolute ruler. The monarchy in Athens and in Rome came before the Republic, in time. All these ambiguities are merely a proof that the various types of State should be defined in some other way.

To find an appropriate definition, let us look back to what has been said of the nature of the State in general. The State, we said, is the organ of social thought. That does not mean that all social thought springs from the State. But there are two kinds. One comes from the collective mass of society and is diffused throughout that mass; it is made up of those sentiments, ideals, beliefs that the society has worked out collectively and with time, and that are strewn in the consciousness of each one. The other is worked out in the special organ called the State or government. The two are closely related. The vaguely diffused sentiments that float about the whole expanse of society affect the decisions made by the State, and conversely, those decisions made by the State, the ideas expounded in the Chamber, the speeches made there and the measures agreed upon by the ministries, all have an echo in the whole of the society and modify the ideas strewn there. Granted that this action and reaction are a reality, there are even so two very different forms of collective psychic life. The one is diffused, the other has a structure and is centralized. The one, because of this diffusion, stays in the half-light of the sub-conscious. We cannot with certainty account for all these collective preconceptions we are subject to from childhood, all these currents of public opinion that form here and there and sway us this way and that. There is nothing deliberately thought out in all this activity. There is something spontaneous, automatic, something unconsidered, about this whole form of life. Deliberation and reflection, on the other hand, are features of all that goes on in the organ of government. This is truly an organ of reflection: although still in a rudimentary stage, it has a future of progressive development. There all is organized and, above all, organized increasingly to prevent changes being made without due consideration. The debates in the assemblies—a process analogous to thought in the individual—have the precise object of keeping minds very clear and forcing them to become aware of the motives that sway them this way or that and to account for what they are doing. There is something childish in the reproaches directed at deliberative assemblies as institutions. They are the sole instruments that the collectivity has to prevent any action that is unconsidered or automatic or blind. Therefore there exists the same contrast between the psychic life diffused throughout society and the parallel life concentrated and worked out especially in governmental organs, as exists between the diffused psychic life of the individual and his clear consciousness. Within every one of us, then, there is at all times a host of ideas, tendencies and habits that act upon us without our knowing exactly how or wherefore. To us they are hardly perceptible and we are unable to make out their differences properly. They lie in the subconscious. They do however affect our conduct and there are even individuals who are moved solely by these motives. But in the part of us that is reflective there is something more. The ego that it is, the conscious personality that it represents, does not allow itself to follow in the wake of all the obscure currents that may form in the depths of our being. We react against these currents; we wish to act with full knowledge of the facts, and it is for this reason that we reflect and deliberate. Thus, in the centre of our consciousness, there is an inner circle upon which we attempt to concentrate light. We are more clearly aware of what is going on there, at least of what is going on in the underlying regions. This central and relatively clear consciousness stands to the nameless and indistinct representations that form the sub-stratum of our mind, as does the scattered collective consciousness of the society to the consciousness lying in the government. Once we have grasped what the special features of this consciousness are and that it is not merely a reflexion of the obscure collective consciousness, the difference between the various forms taken by the State is easily recognized.

And so we perceive that this government consciousness may be concentrated in the organs that have rather limited scope, or again, may be spread through the society as a whole. Where the government organ is jealously guarded from the eyes of the many, all that happens within it remains unknown. The dense mass of society receives the effect of its actions without taking part, even at a distance, in its discussions and without perceiving the motives that decide those who govern on the measures they decree. As a result, what we have called the government consciousness remains strictly localized in these particular spheres, that are never very extensive. But it sometimes happens that these, as it were, watertight bulkheads that separate this particular *milieu* from the rest of the society are less impervious. It does occur that a great deal of the action taken in this *milieu* is done in the full light of day and that the debates there may be so conducted as to be heard by all. Then, everyone is able to realize the problems set and the circumstances of the setting and the at least apparent reasons that determine the decisions made. In this case, the ideas, sentiments, decisions, worked out within the governmental organs do not remain locked away there; this whole psychic life, so long as it frees itself, has a chain of reactions throughout the country. Every one is thus able to share in this consciousness *sui generis* and asks himself the questions those governing ask themselves; every one ponders them, or is able to. Then, by a natural reversal, all the scattered reflections that ensue in this way, react on the governmental thought which was their source. From the moment that the people set themselves the same questions as the State, the State, in solving them, can no longer disregard what the people are thinking. It must be taken into account. Hence the need for a measure of consultation, regular or periodic. It is not because the custom of such consultations had become established that governmental life was communicated the more to the citizens, taken as a whole. It is because such communication had previously become established of itself that these consultations became imperative. And the fact that has given rise to such communication is that the State has ceased more and more to be what it was over a long era; that is, a kind of mysterious being to whom the ordinary man dared not lift his eyes and whom he even, more often than not, represented to himself as a religious symbol. The representatives of the State bore the stamp of a sacred character and, as such, were set apart from the commonalty. But by the gradual flow of ideas the State has little by little lost this kind of transcendence that isolated it within itself. It drew nearer to men and men came to meet it. Communications became closer and thus, by degrees, this circuit—just described—was set up. The governmental power, instead of remaining withdrawn within itself, penetrated down into the deep layers of the society, there received a new turn of elaboration and returned to its point of departure. All that happens in the *milieux* called political is observed and checked by every one, and the result of this observing and checking and of the reflections they provoke, reacts on the government *milieux*. By these signs we recognize one of the distinctive features of what is usually called democracy.

We must therefore not say that democracy is the political form of a society governing itself, in which the government is spread throughout the *milieu* of the nation. Such a definition is a contradiction in terms. It would be almost as if we said that democracy is a political society without a State. In fact, the State is nothing if it is not an organ distinct from the rest of society. If the State is everywhere, it is nowhere. The State comes into existence by a process of concentration that detaches a certain group of individuals from the collective mass. In that group the social thought is subjected to elaboration of a special kind and reaches a very high degree of clarity. Where there is no such concentration and where the social thought remains entirely diffused, it also remains obscure and the distinctive feature of the political society will be lacking. Nevertheless, communications between this especial organ and the other social organs may be either close or less close, either continuous or intermittent. Certainly in this respect there can only be differences of degree. There is no State with such absolute power that those governing will sever all contact with the mass of its subjects. Still, the differences of degree may be of significance, and they increase in the exterior sense with the existence or non-existence of certain institutions intended to establish the contact, or according to the institutions' being either more or less rudimentary or more or less developed in character.

It is these institutions that enable the people to follow the working of government (national assembly—parliament, official journals, education intended to equip the citizen to one day carry out his duties—and so on . . .) and also to communicate the result of their reflections (organ for rights of franchise or electoral machinery) to the organs of government, directly or indirectly. But what we have to decline at all costs is to admit a concept which (by eliminating the State entirely) opens a wide door to criticism. In this sense, democracy is just what we see when societies were first taking shape. If every one is to govern, it means in fact that there is no government. It is collective sentiments, diffused, vague and obscure as they may be, that sway the people. No clear thought of any kind governs the life of peoples. Societies of this description are like individuals whose actions are prompted by routine alone and by preconception. This means they could not be put forward as representing a definite stage in progress: rather, they are a starting point. If we agree to reserve the name democracy for political societies, it must not be applied to tribes without definite form, which so far have no claim to being a State and are not political societies. The difference, then, is quite wide, in spite of apparent likeness. It is true that in both cases—and this gives the likeness—the whole society takes part in public life but they do this in very different ways. The difference lies in the fact that in one case there is a State and in the other there is none.

This primary feature, however, is not enough. There is another, inseparable from it. In societies where it is narrowly localised, the government consciousness has, too, only a limited number of objects within its range. This part of public consciousness that is clear is entirely enclosed within a little group of individuals and it is in itself also only of small compass. There are all sorts of customs, traditions and rules which work automatically without the State itself being aware of it and which therefore are beyond its action. In a society such as the monarchy of the seventeenth century the number of things on which government deliberations have any bearing is very small. The whole question of religion was outside its province and along with religion, every kind of collective prejudice and bias: any absolute power would soon have come to grief if it

83

had attempted to destroy them. Nowadays, on the other hand, we do not admit there is anything in public organization lying beyond the arm of the State. In principle, we lay down that everything may for ever remain open to question, that everything may be examined, and that in so far as decisions have to be taken, we are not tied to the past. The State has really a far greater sphere of influence nowadays than in other times, because the sphere of the clear consciousness has widened. All those obscure sentiments which are diffusive by nature, the many habits acquired, resist any change precisely because they are obscure. What cannot be seen is not easily modified. All these states of mind shift, steal away, cannot be grasped, precisely because they are in the shadows. On the other hand, the more the light penetrates the depths of social life, the more can changes be introduced. This is why those of cultivated mind, who are conscious of themselves, can change more easily and more profoundly than those of uncultivated mind. Then there is another feature of democratic societies. They are more malleable and more flexible, and this advantage they owe to the government consciousness, that in widening has come to hold more and more objects. By the same token, resistance is far more sharply defined in societies that have been unorganized from the start, or pseudo-democracies. They have wholly yielded to the yoke of tradition. Switzerland, and the Scandinavian countries, too, are a good example of this resistance.

To sum up, there is not, strictly speaking, any inherent difference between the various forms of government; but they all lie intermediate between two contrasting schemes. At one extreme, the government consciousness is as isolated as possible from the rest of the society and has a minimum range

A difficulty comes perhaps in distinguishing between the two kinds of society, aristocratic and monarchic. The closer communication becomes between the government consciousness and the rest of society, and the more this consciousness expands and the more things it takes in, the more democratic the character of the society will be. The concept of democracy is best seen in the extension of this consciousness to its maximum and it is this process that determines the communication.

VIII

CIVIC MORALS (Continued)

FORM OF THE STATE—DEMOCRACY

IN the last lecture we saw that it was quite impossible to define a democracy and other forms of the State according to the number of those governing. Except for small tribes of the lowest order there are no societies where the government is carried out direct by all in common: it is always in the hands of a minority chosen either by birth or by election; its scope may be large or small, according to circumstances, but it never comprises more than a limited circle of individuals. In this respect there are only slight shades of difference between the various political forms. Governing remains always the function of an organization that is clearly defined and hence has definite limits. But what does vary appreciably, according to the society, is the way in which the government organ communicates with the rest of the nation. Sometimes contact is at long intervals or irregular; the government keeps itself out of sight and lives retired within itself; at other times, it keeps in contact with the society only fitfully and fails to reach all parts of it. It is not constantly aware of the society and the society is not aware of it. We may ask what use it does make of its activity in the circumstances? For the main part it is turned towards the outer world. It takes little part in the internal life, because its own life is elsewhere: it is above all the agent of external relations, the agent for the acquisition of territory and the organ of diplomacy. In other societies, by contrast, communications between the State and other parts of society are many, and both regular and organized. The citizens are kept in touch with what the State is doing and the State is at given periods, or it may be continuously, told of what is going on in the deep layers of the society. It is informed either through administrative channels or by the voice of the electorate about what is happening in the most distant or most obscure strata of the society, and these in turn are informed as to the events going on in political circles. The citizens take part from a distance in some of the deliberations that go on there; they are aware of the action taken, and their judgments and the result of their considered thought comes back to the State by certain channels. This is really the gist of democracy. It is of little moment that those who hold the reins of State run to this number or that; what is essential and a special feature is the way in which they communicate with the society as a whole. It is true that, even in this respect, there are only differences of degree between the various types of political system, but the differences are in this case deeply marked; further, they can be recognized externally by the existence or otherwise, of institutions for ensuring that close communication which is a feature of the democratic form.

This primary characteristic, however, is not the only one: there is another, bound up with the first. The more government consciousness is localized within the strict limits of the organ, the fewer the number of objects this consciousness affects. The fewer the ties that link it with the various regions of society, the smaller its compass. This is quite natural, for whence could it draw any sustenance, seeing that its dealings with the rest of the nation are only distant and fitful. The organ of government is only faintly conscious of what is going on in the interior of the organ-society, and therefore, by the force of circumstance, nearly the whole of collective life remains obscure, diffused and unconscious. It is made up entirely of uncharted traditions, of prejudices, of obscure sentiments that no organ could get hold of to bring into the light. Compare the small number of things that government deliberations covered in the seventeenth century and the thousand-and-one objects they apply to nowadays. The difference is vast. Formerly, public activity was concerned almost solely with external matters. The whole of the law worked automatically in an unconscious way; it was a matter of custom. It was the same with religion, education, health and economic life—to a great extent, at any rate. Local and regional interests were left to themselves and ignored. Nowadays, in a State like our own (and, allowing for difference in degree, increasingly so in the great European States), all that is involved in the administration of justice, in education, in the economic life of the people, has become conscious. Each day brings debates on these questions that arouse different reactions. And this difference is even perceptible to those outside that sphere. Whatever is diffuse, obscure, unrecognizable, escapes our action. If the character of this obscure un-named is not known or barely known, it is hardly possible to change it. If ideas or sentiments are to be modified, they must first be brought into view and grasped as clearly as possible and their nature realized. This explains why the more an individual is conscious of himself and able to reflect, the more accessible he is to change. Uncultivated minds, on the other hand, are the rigid minds of routine, in which no idea can take hold. For the same reason, when collective ideas and sentiments are obscure or unconscious, when they are scattered piecemeal throughout the society, they resist any change. They elude any action because they elude consciousness: they cannot be grasped because they are in the shadows. The government cannot affect them at all. Further, it is an error to believe that governments we term absolute are all-powerful. It is one of those illusions that come from a superficial view. They are indeed all-powerful against the individual and this is what the term of 'absolute' means, as applied to them; in that sense it is justified. But against the social condition (*état*) itself, against the structure of society, they are relatively powerless. Louis XIV, clearly, was able to issue his *lettres de cachet* against anyone he wished, but he had no power to modify the existing laws and usages, the established customs or accepted beliefs. What could avail him against the Church, and its manifold privileges that enabled that organization to defy any government action? The privileges of town and guild resisted all efforts made to modify them, up to the end of the *ancien régime*. We know, too, with how slow a pace the law was developing at that time. We have only to compare how rapidly important changes come about to-day in various fields of social activity. Every day some new law is put on the statute-book, another is taken off or some new modification is made in religious or government organization, or in education, and so on. . . . That is because all these obscure things come more and more to the surface in that region of the social consciousness that is lucid, which is the government consciousness. As a result, it becomes all the more malleable. The clearer ideas and sentiments become, the more completely they are dominated by reflection and the greater its hold on them. That is to say, they can be freely criticized and debated, and these discussions have the inevitable effect of making them lose their powers of resistance, of making them more accessible to change, or even of changing them direct. Here again, in this enlargement of the field of government consciousness and in this greater malleability, we see yet another distinguishing feature of democracy. Just because a greater number of things is submitted to collective debate there are, too, more things on the road to becoming reality. Traditionalism, on the contrary, is a feature of the other political forms. Here again, the distinction is very clear in respect of pseudo-democracies, examples of which can be found in the lower societies and which, again on the contrary, are incapable of shedding custom and tradition.

To sum up : if we want to get a fairly definite idea of what a democracy is, we must begin by getting away from a number of present concepts that can only muddle our ideas. The number of those governing must be left out of account and, even more important, their official titles. Neither must we believe that a democracy is necessarily a society in which the powers of the State are weak. A State may be democratic and still have a strong organization. The true characteristics are twofold : (1) a greater range of the government consciousness, and (2) closer communications between this consciousness and the mass of individual consciousnesses. The confusions that have occurred can be understood to some extent by the fact that in societies where the government authority is weak and limited, the communications linking it to the rest of the society are of necessity quite close, since it is not distinguishable from the rest. It has no existence outside the mass of the people, it must therefore of necessity be in communication with that mass. In a small primitive tribe, the political leaders are only delegates and always provisional, without any clearly defined functions. They live the life of everyone else, and their decisive discussions remain subject to the check of the whole collectivity. They do not however form a separate and definite organ. And here we find nothing resembling the second feature already mentioned —I mean the pliability deriving from the range of government consciousness, that is, from the field of collective, clear ideas.

Societies such as these are the victims of traditional routine. This secondary feature is perhaps even more distinctive than the first. The first criterion, at any rate, can be very useful providing it is used with discernment, and providing we beware of identifying the confused situation arising from the State not yet being detached from the society and separately organized, with the communications that may exist between a clearly defined State and the society it governs.

Seen from this point, a democracy may, then, appear as the political system by which the society can achieve a consciousness of itself in its purest form. The more that deliberation and reflection and a critical spirit play a considerable part in the course of public affairs, the more democratic the nation. It is the less democratic when lack of consciousness, uncharted customs, the obscure sentiments and prejudices that evade investigation, predominate. This means that democracy is not a discovery or a revival in our own century. It is the form that societies are assuming to an increasing degree. If we can once do without the ordinary labels that only vitiate clear thinking, we shall admit that seventeenth century society was more democratic than that of the sixteenth and more democratic than any society having a feudal basis. With feudalism, there is diffusion of social life, and obscurity and lack of consciousness are at their worst: this is just what large-scale societies of the present time have succeeded in keeping within bounds. The monarchy, by concentrating the collective forces to an increasing degree, by spreading its roots in all directions and by permeating the social masses more intensely, prepared for a future democracy and was itself—in comparison with what preceded the monarchy—a democratic form of government. It is quite a minor fact that the Head of State then bore the name of king; what we have to weigh are the relations he maintained with the country as a whole. It was the country, from that time onwards, that was itself responsible for the clearsightedness of social ideas. So that it is not just for the last forty or fifty years that democracy has been flowing at high water; the tide has been rising from the beginning of history.

The determining element in this progress is easy to understand. The more societies grow in scope and complexity the more they need reflection in conducting their affairs. Blind routine and a uniform tradition are useless in running a mechanism of any delicacy. The more complex the social *milieu* becomes, the greater its mobility. The social structure has to change on the same scale and to achieve this it has to be conscious of itself and capable of reflection. When things go on happening in the same way, habit will suffice for conduct; but when circumstances are changing continually, habit, on the contrary, must not be in sovereign control. Reflection alone makes possible the discovery of new and effectual practices, for it is only by reflection that the future can be anticipated. This is why deliberative assemblies are becoming ever more widely accepted as an institution. They are the means by which societies can give considered thought to themselves, and therefore they become the instrument of the almost continuous changes that present-day conditions of collective existence demand. The social organs, if they are to survive, must be ready for change. If they are to change in good time and rapidly at that, the reflective powers of the society have to follow the course of changing circumstances and organize, too, the means of adapting themselves to the changes. Whilst the advances in democracy are made thus inevitable by the state of the social *milieu*, they are prompted equally by our inmost moral concepts. Democracy indeed, as we have defined it, is the political system that conforms best to our present-day notion of the individual. The values we attribute to individual personality make us loth to use it as a mechanism to be wielded from without by the social authority. The personality can be itself only to the degree in which it is a social entity that is autonomous in action. It is true that in one sense it receives everything from without—its moral and its material energies. Just as we can sustain our physical life only by the aid of sustenance taken from the cosmic *milieu*, so do we give sustenance to our mental life only by the aid of ideas and sentiments that reach us from the social *milieu*. Nothing yields nothing, and the individual left to his own devices could not raise himself beyond his own level. What makes it possible for him to transcend himself and to rise above the level of animal nature is, that collective life echoes in him and permeates him; it is these adventitious elements that give him a different nature. But there are two ways in which a human being can receive help from exterior forces. Either he receives them passively, unconsciously, without knowing why—and in this case, he is only a thing. Or, he is aware of what they are, of his reasons for submitting and being receptive to them; in that case he is not passive, he acts consciously and of his own accord, knowing well what he is about. The action is in this sense only a passive state, whose *raison d'être* we know and understand. The autonomy an individual can enjoy does not, then, consist in rebelling against nature—such a revolt being futile and fruitless, whether attempted against the forces of the material world or those of the social world. To be autonomous means, for the human being, to understand the necessities he has to bow to and accept them with full knowledge of the facts. Nothing that we do can make the laws of things other than they are, but we free ourselves of them in thinking them, that is, in making them ours by thought. This is what gives democracy a moral superiority. Because it is a system based on reflection, it allows the citizen to accept the laws of the country with more intelligence and thus less passively. Because there is a constant flow of communication between themselves and the State, the State is for individuals no longer like an exterior force that imparts a wholly mechanical impetus to them. Owing to constant exchanges between them and the State, its life becomes linked with theirs, just as their life does with that of the State.

Even so, there does exist nevertheless a concept of democracy and its practice which must be carefully distinguished from the one just discussed.

It is often said that under a democratic system the will and thought of those governing are identical and merge with the will and thought of those governed. From this standpoint the State does no more than represent the mass of the individuals, and the whole governmental structure can have only the aim of transmitting as faithfully as possible the sentiments diffused throughout the collectivity, with nothing added and nothing modified. The ideal, we might say, would consist in expressing them as adequately as possible. It is to this concept that the usage of what is called the 'imperative mandate' of the electorate, as well as of its substitutes, clearly corresponds. For although this mandate has not become part of our morals in its pure form, the ideas that form its basis are very widespread.

This way of forming an image of those governing and their functions is fairly common. But nothing can be more contrary, in some respects, to the very notion of democracy. For democracy pre-supposes a State, an organ of government, distinct from the rest of the society, although closely in contact : so that this kind of view is the very negation of any State in the true sense of the term, since it re-absorbs the State in the nation. If the State does no more than receive individual ideas and volitions to find out which are the most widespread and 'in the majority', as it is called, it can bring no contribution truly its own to the life of society. It is only an offprint of what goes on in the underlying regions. It is this that stands in contradiction to the very definition of the State. The role of the State, in fact, is not to express and sum up the unreflective thought of the mass of the people but to superimpose on this unreflective thought a more considered thought, which therefore cannot be other than different. It is and must be a centre of new and original representations which ought to put the society in a position to conduct itself with greater intelligence than when it is swayed merely by vague sentiments working on it. All these deliberations, all these discussions, all this enquiry by statistics, all this administrative information put at the disposal of government councils, and which go on increasing in volume—all these are the starting point of a new mental life. Thus, material is collected which is not available to the mass of the people and it undergoes a process of elaboration of which this mass is not capable, because it has no unity, is not gathered within one enclosure, and its attention cannot be applied at the same moment to the same object. Is it not inevitable that something new must emerge from all this activity? The duty of the government is to make use of all these means, not simply so as to arrive at what the society is thinking but to discover what is in its best interests. It is better placed than the crowd to know what is expedient: it must therefore see things in a different light from the crowd. It is true it has to be informed as to what the citizens are thinking, but this is only one of the elements in its means of deliberation and reflection, and since it is framed to think along special lines, it has to take thought in its own way. That is the *raison d'être* of government. Equally, it is essential that the rest of the society should know what it is about,

what its thought is and should be able to follow this in forming
a judgment: it is necessary that there should be as complete a
harmony as possible between both these parts of the social
structure. But this harmony does not imply that the State should
be in servitude to the citizens and reduced to being a mere
echo of their will. Such a concept of the State is too obviously
close to the concept lying at the base of the so-called primitive
democracies. Here the distinction lies in the exterior structure
of the State being complex and skilfully devised in a different
sense. We could not compare a council of sachems to our own
government organization, even though its function were obviously the same. There would be no autonomy in either case.
What would follow? Such a State would fail in its mission:
instead of giving clarity to the vague sentiments of the mass of
the people and subordinating them to clearer and more
reasoned ideas, it would only allow those sentiments to prevail
which seemed to have the most general currency.

This is not the only drawback attached to such a concept.
We have seen that in the lower societies, the lack of any government or the rudimentary character of a weak one, result in a
relentless traditionalism. This is so because the society has
strong and vigorous traditions, deeply engraved in the individual consciousnesses, and these traditions are powerful precisely
because the societies are simple in form. But it is not the same in
the large-scale societies of the present day; traditions have lost
their sway, and as they are incompatible with the spirit of
scrutiny and free criticism, the acute need of which is always
growing, they cannot and should not maintain their earlier
authority. And what is the result? We find it is the individuals
who (within the concept of democracy under review), give
the motive power to those governing: the State is incapable of
bringing a moderating influence to bear on them. On the other
hand, they do not any longer find within themselves a sufficing
number of deep-rooted ideas and sentiments as would keep
them from giving way to the first gusts of doubt or debate.
There are not many left of those despotic States strong enough
to stand above all criticism, nor of those beliefs and practices
that are beyond argument. Therefore the citizens are not restrained from without by the government, because it follows
in their wake, nor from within by the state of the ideas and

collective sentiments they harbour: so that all, in practice as in theory, becomes a matter of controversy and division and all is in a state of vacillation. There is no firm ground under the feet of the society. Nothing any longer is steadfast. And since the critical spirit is well developed and everyone has his own way of thinking, the state of disorder is made even greater by all these individual diversities. Hence the chaos seen in certain democracies, their constant flux and instability. There we get an existence subject to sudden squalls, disjointed, halting and exhausting. If only this state of affairs led to any really profound changes. But those that do come about are often superficial. For great changes need time and reflection and call for sustained effort. It often happens that all these day-to-day modifications cancel each other out and that in the end the State remains utterly stationary. These societies that are so stormy on the surface are often bound to routine.

It would be useless to hide from ourselves that this situation applies in part to our own country. These ideas, according to which the government is only the transmitter of general volitions, are current too with us. They lie, too, at the base of Rousseau's doctrine; with reservations that may be of varying significance they are, again, at the base of our own parliamentary practice. So it is of extreme importance to know their causes.

No doubt it would be easy to tell ourselves that they are simply due to an intellectual fallacy and that they amount to a simple error in logic; that to correct this it would be enough to point it out, to demonstrate it with evidence and to prevent it going on by using education and the right kind of warnings. But collective errors, like individual ones, derive from objective causes and can only be cured by tackling the causes. If individuals affected by colour-blindness make mistakes about colours it is because the organ of sight is formed in such a way as to cause this failure, and no matter how we may warn them, they will go on seeing things as they see them. Likewise, if a nation has a certain way of representing to itself the role of the State and the nature of its relations with the State, that is because there is something in the state of society that makes this false representation inevitable. And all the admonishing and exhorting in the world will not succeed in dispelling it, so long as the organic constitution that decides it is not modified. True, it is not altogether fruitless to let someone sick know the nature of the illness and its handicaps, but to cure him we still have to make him see all the factors in it so that he may deal with them and be cured. No changes will come about with fine words alone.

In the case in question, it seems inevitable that this variant of democracy should replace the normal form whenever the State and the mass of individuals are in direct relation, without any intermediary being brought in between them. For as a result of this proximity, it is necessary as a point of mechanics that the weakest collective force, that is, the State, should not be absorbed by the strongest, that of the nation. When the State stands too close to the individuals, it falls under their dominance and at the same time is in their way. Its proximity is a hindrance, for even so, it attempts to control them direct by rule when, as we know, it is incapable of playing this part. But that same proximity makes it depend closely on them, for the individuals, being numerous, can change the State at will.

From the moment that we have the citizens electing their representatives direct, that is, electing those members with most influence in the governmental organ, it becomes inevitable for these representatives to apply themselves almost exclusively to a faithful promotion of their constituents' views. It is also inevitable for these constituents to claim this docile attitude as an obligation. Does it not amount to a mandate negotiated between the two parties? True, it might be in the nature of higher policy to consider that those governing should enjoy a good deal of initiative and that it is only on those terms that they can carry out their given task. But there is a force of circumstances against which even the best reasoning cannot prevail. As long as the political order places the deputies and, as a rule, governments, in immediate contact with the mass of citizens, it is practically impossible that these latter should not make the laws. This is why acute thinkers have often claimed that members of the political assemblies should be elected by vote in two or more stages. This is because the intermediaries that are intercalated set free the hands of the government; and such intermediaries have been introduced without causing

95

any break in contacts which any government council may have with another. It is not at all imperative that the contacts should be direct. Life must circulate without a break in continuity between the State and individuals and vice versa; but there is no reason whatever why this circulation should not be by way of agencies that are introduced. By means of this intercalation the State will be more dependent on itself, the distinction between it and the rest of the society will be clearer, and by that very fact it will be more capable of autonomy.

So that our political malaise is due to the same cause as our social malaise: that is, to the lack of secondary cadres to interpose between the individual and the State. We have seen that these secondary groups are essential if the State is not to oppress the individual: they are also necessary if the State is to be sufficiently free of the individual. And indeed we can imagine this as suiting both sides; for both have an interest in the two forces not being in immediate contact although they must be linked one with the other.

But what are the groups which are to free the State from the individual? Those able to fulfil this are of two kinds. First, the regional groups. We could imagine, in fact, that the representatives of the *communes* of one and the same *arrondissement*, perhaps even of one and the same *département*, might constitute the electoral body having the duty of electing the members of the political assemblies. Or professional groups, once set up, might be of use for this task. The councils with the duty of administering each of these groups would nominate those who would govern the State. In both cases there would be continuous communication between the State and its citizens, but no longer direct. Of these two methods of organization, one would seem to be more suited to the general orientation of our whole social development. It is quite certain the regional districts have not the same importance as they once had, nor do they any longer play the same vital role. The ties which unite members of the same *commune* or the same *département* are fairly external. They are made and unmade with the greatest ease since the population has become so mobile. There is therefore something rather exterior and artificial about such groups. The permanent groups, those to which the individual devotes his whole life, those for which he has the strongest attachment, are the professional groups. It therefore seems indeed that it is they which may be called upon to become the basis of our political representation as well as of our social structure in the future.

IX

CIVIC MORALS (End)

FORM OF THE STATE—DEMOCRACY

HAVING defined the nature of democracy, we have seen that the concept and practice of it may be of a kind to change that nature seriously for the worse. In essence, it is a system where the State, whilst remaining distinct from the mass of the nation, is closely in communication with it and where its activity therefore reaches some degree of mobility. Now, we have seen that in some cases, this close communication may go so far as to be an almost complete fusion. Instead of being a well-defined organ, the centre of an original life of its own, the State then becomes merely an offprint of the life underlying it. It does no more than translate what individuals think and feel, in a different notation. Its role is no longer that of elaborating new ideas and new points of view—a task for which its framework fits it. No, its main functions consist of reckoning what the ideas and sentiments are that have the widest circulation, those of 'the majority', as they say. The State is the result of this very reckoning. The election of deputies simply means counting the supporters of certain opinions in the country. Such a concept is, however, contrary to the idea of a democratic State, since it eliminates almost entirely the very idea of the State. I say 'almost entirely', for the fusion is of course never complete. The very force of circumstances makes it impossible for the mandate given to a deputy to be framed in such definite terms as to bind him completely. Some slight lee-way of initiative must always remain. But at any rate there is the tendency to reduce that lee-way. It is in this sense that any such political system approximates to what we observe in primitive societies, for in both cases the governmental power is weak. But there is this vast difference, all the same, that in the one case the State does not yet exist, or exists only in embryo, whilst in this variant of democracy it is,

98

on the contrary, quite often very far developed, with an extensive and complex structure. And it is just this twofold, contradictory aspect that best shows the abnormal character of the phenomenon. On the one hand we have a mechanism that is complex and ingenious, the multiple cogwheels of a vast administration; on the other, a concept of the part played by the State that represents a return to the most primitive of political forms. Hence, a strange mixture of inertia and activity. The State does not move of its own power, it has to follow in the wake of the obscure sentiments of the multitude. At the same time, however, the powerful means of action it possesses makes it capable of a heavy hand on the same individuals whose servant, otherwise, it still remains.

We have already said that this view of the concept and practice of democracy was still deeply rooted in French minds. Rousseau, whose philosophy put these ideas into systematic form, remains the devisor of our democratic theory. Indeed, there is no political philosophy which offers a better example of this dual, paradoxical aspect just described. If we look at it in one way, it is narrowly individualistic—the individual is the moving principle of the society, the society being only the sum total of individuals. We know on the other hand what authority Rousseau assigns to the State. Further, the proof that these ideas are actively current with us is seen in the whole spectacle of our political life. We cannot deny that, seen from without, and on the surface, it appears to have altogether excessive mutability. Change follows on change with unparalleled speed. It is many years now, since it has succeeded in steering any fixed course for long. As we have seen, that was bound to be the case from the moment that the driving force of the State came from a multitude of individuals who regulated its course with almost supreme power. At the same time, these surface changes mask an habitual stagnation. We must deplore the constant flux in political events and the all-powerful government offices with their inveterate clinging to tradition. They are a force against which we are powerless. This is because all these surface changes that go on in various directions cancel each other out: nothing remains except the fatigue and exhaustion that are a feature of these unceasing mutations. The result is that habits strongly entrenched and the routines that remain untouched by these

99

changes, have all the greater sway, for they alone are effectual. Their power derives from the excessively fluid state of the rest. And we do not know whether to deplore this or to welcome it: for there is always a residue of organization that holds, a little stability and resolve, where they are needed for survival. Despite all its defects, it is quite possible the administrative machine is giving very valuable service at this time.

We have diagnosed the evil, but what is its source? It is a false concept, but false concepts have objective causes. There must be some element in our political constitution to explain this error.

What seems to have produced the error is the special character of our present structure, by virtue of which the State and the mass of individuals stand in direct contact and communication without any intermediary. The electorate is made up of the whole enfranchised population of the country and it is from it that the State derives—at least the vital organ of the State, that is, the deliberative assembly. So it is inevitable that a State made up in these conditions be, more or less, simply a reflection of the social mass and nothing more. Here we are confronted by two collective forces—one of vast proportions, made up as it is of all citizens together, the other extremely limited, because it includes only the representatives. So, by a physical law, it is inevitable that the weaker should follow in the wake of the greater. From the moment we have the individuals electing their representatives direct, the deputies are bound to confine themselves exclusively to faithfully interpreting the wishes of their constituents, and for these to claim their docility as an obligation. It is true it would be in the nature of higher policy to consider that those governing should enjoy a good deal of initiative, and that it is only on these terms they could carry out their task; that in the common interest they should see things differently from the individual—taken up as he is with his other social functions—and that therefore the State should be allowed to act according to its nature. But there is a force of circumstances against which even the best reasoning cannot prevail. As long as the political order brings the deputies in immediate contact with the unorganized mass of individuals, it is inevitable that the latter should make the laws. This direct contact does not allow the State to be itself.

This is why some thinkers demand that members of the political assemblies should be nominated by vote in two or more stages. It is indeed quite certain that the only means of releasing the government is to devise intermediaries between it and the rest of the society. It is true there must be continuous communication between government and all the other social organs, but this must not go so far as to make the State lose its identity. The State must have relationship with the nation without being absorbed in it, and therefore they must not be in immediate contact. The only means of preventing a lesser force from falling within the orbit of a stronger is to intercalate between the two, some resistant bodies which will temper the action that has the greater force. From the moment that the State emerges less immediately from the mass of the people, so much the less is it subject to the action of that mass: it is able to belong to itself all the more. Obscure tendencies, dimly at work in the country, no longer exercise the same influence over its efforts and no longer have the effect of curbing its decisions so closely. This result, however, can only be fully reached if the groupings thus intercalated between the greater number of the citizens and the State are natural and enduring. It is not enough, as sometimes believed, to simply intercalate artificial intermediaries, created solely for the occasion. We might, for instance, be satisfied to set up (one by one, by means of electoral bodies comprising the sum total of the electorate), a more limited body which—either direct or by the intermediary of a still smaller body—would nominate those to govern. This body, its task once carried out, would pass out of existence and the State, set up by these means, might well enjoy a certain independence; but it would no longer adequately fulfil the other condition that goes with a democracy: it would no longer be in close communication with the country as a whole. For once it had come to life, and the main intermediary and the lesser ones which had served to form it had ceased to exist, there would be a vacuum between it and the mass of the people. There would not be those constant exchanges between one and the other which are so essential. It is important that the State should not be under the dominance of individuals, but that does not mean it should ever lose contact with them. This inadequate communication with the people as a whole is what makes for the weakness of any deliberative assembly recruited in this way. This would be too out of touch with popular needs and sentiments, which cannot be brought into its notice with proper continuity. Thus, one of the elements vital to its deliberations is lacking.

In order that contact should never be lost, the intermediary bodies thus intercalated should not be merely set up temporarily but should remain continuously in operation. In other words, they must be natural and normal organs of the social body. There are two kinds of intermediary which could play this part. First, the secondary councils that have charge of administering regional areas. We can imagine, for instance, that the councils of the *départements* or provinces, whether elected by direct or indirect vote, might be called upon to take over this function. It would be for them to nominate members of the government councils and of deliberative assemblies that are genuinely political.

It is roughly this idea that served as the basis for organizing our present-day Senate. What makes us doubt whether such an arrangement would be the best suited to the constitution of the great European States, is that regional sub-divisions of countries are losing their significance to an increasing extent. As long as every district, *commune*, area or province had its own peculiar features, its traditional morals, its customs and its special interests, the councils set up to administer them were essential cogwheels in political life. It was in those councils that the ideas and aspirations that stir the masses concentrated direct, without any medium. But nowadays, the links that bind each one of us to a particular spot in an area where we live are incalculably frail and can be broken with the greatest ease. We are here one day and elsewhere the next. We feel as much at home in one province as another, or at least, the special affinities of a regional origin are quite secondary and no longer have any great influence on our life. Even though we remain attached to the same place, our interests go far and away beyond the administrative limits of the area where we happen to be living. The way of life immediately surrounding us is not even the life that is of the deepest concern to us. Whatever I may be—professor, manufacturer, engineer or artist—it is not the events that happen in my own *commune* or *département* that concern me most directly and excite me. I can even live my every-day life and know nothing of them at all.

What matters far more to us, according to the functions we have to fulfil, is what goes on at scientific conferences, what is being published, what is being said in the great centres of production; the new events in the world of art in the big cities of France or abroad have an interest that is very different for the painter or sculptor from, say, municipal affairs. We might say the same of the manufacturer who has connexions with all sorts of industries and trading concerns throughout his own region and even far afield in the world. No one will deny that the grouping that is merely regional is rapidly declining. But then, the councils in charge of administering these groups are not able to concentrate and give expression to the general life of the country, for the way in which this life is dispersed and organized does not, as a rule at least, reflect the regional division of the country. And what is the reason that these councils lose their standing, why is there less canvassing for the honour of a seat on them, and why do the enterprising minds, the men of talent, seek a different field of activity? It is because they are organs that are indeed rather on the wane. A political assembly resting on such a basis can only give an imperfect picture of a society's structure or of the relation between the various social forces.

Professional life, on the other hand, takes on increasing importance, as labour goes on splitting up into divisions. There is therefore reason to believe that it is this professional life that is destined to form the basis of our political structure. The idea is already gaining ground that the professional association is the true electoral unit, and because the links attaching us to one another derive from our calling rather than from any regional bonds of loyalty, it is natural that the political structure should reflect the way in which we ourselves form into groups of our own accord. Let us imagine the guilds or corporative bodies established or revived according to the plan we have outlined: each would have at its head a council to direct it and govern its internal affairs. Would these various councils not be wonderfully appropriate to play this part of intermediary electoral units, which are at present served only feebly by the regional groups? Professional life is never interrupted and is never at a standstill. The corporative body or guild and its organs are always in action and therefore the governmental assemblies that would issue from them would never lose touch with the councils of a society: they would, too, never run the risk of being isolated within themselves, or of not feeling quickly and vividly enough the changes that happen to occur in the deep-lying strata of the population. Independence would be ensured without any interruption in communication.

Such a combination would also have two other advantages worth noting. Universal suffrage, as it is in practice to-day, has often been blamed for being inadequate for its purpose. It is said, not unjustly, that a deputy could never be armed with enough facts to settle the countless questions laid before him. But this incompetence of the deputy only reflects that of the elector and this inadequacy is the more serious thing. For since the deputy has a mandate from the electors and is expressly charged with conveying the views of those he represents, these individuals must equally consider the same problems and thus themselves assume the very same general competence. In fact, whenever consulted, the elector makes up his mind on all vital questions that arise in the deliberative assemblies and the election takes the form of a numerical return of all the individual opinions that find expression in this way. We need not stress the fact that they could not be well-informed. It would be different if the voting were organized on the basis of corporative bodies. As far as the interests of the professions are concerned, every active member there is competent in his own line. It would, then, be most appropriate to choose those who are best at conducting the general business of the corporative body. Moreover, the delegate sent by these people to the political assemblies would go armed with their own particular proficiencies; and since the main task of these assemblies would be to regulate inter-professional relations, they would be made up in the most suitable way for solving such problems. The councils of government would then be genuinely what the brain is to the human organism—a reflexion of the social body. All the living forces, all the vital organs would be represented there according to their relative importance, In the group thus formed, the society would truly gain consciousness of itself and of its unity; this unity would follow naturally from the relations that would develop amongst those representing the different professions thus placed in close contact.

In the second place, there is one defect inherent in the framing of any democratic State. Since individuals alone form the living, active substance of society, it follows that the State, in one sense, can be the business only of individuals; in spite of this, the State has to give expression to something quite different from individual sentiments. The State must derive from the individuals and at the same time it has to go beyond them. How then is this paradox to be resolved, which Rousseau in vain wrestled with? The only way of making anything more of individuals than merely themselves is to put them in contact and to group them in a lasting way. The only sentiments rising above individual feelings are those that come about from actions and reactions amongst individuals in association. Let us apply the idea to political organization. If each individual, independently, comes along with his vote to set up the State or the organs which are to serve in giving it definite form; if each one makes his choice in isolation, it is almost impossible for such votes to be inspired by anything except personal and egotistic motives: these will predominate, at any rate, and an individualistic particularism will lie at the base of the whole structure. But let us suppose that such nominations were made as a result of long, collective preparation, and their character would be quite different. For when men think in common, their thought is partly the work of the community. It acts upon them, weighs upon them with all its authority, restraining egotistic whims and setting minds on a common course. Therefore if votes are to be an expression of something more than individuals and if they are to be animated by a collective mind, the ordinary voting electorate should not be made up of individuals brought together solely for this exceptional occasion; they do not know one another, they have not contributed to forming each other's opinions and they merely go along in single file to the ballot box. No, on the contrary, it must be an established group that has cohesion and permanence, that does not just take shape for the moment on polling day. The guild or corporative body corresponds clearly to this desired end. The sentiments of the members who form it are evolved in common and express the community because they are constantly and closely in contact.

Our political malaise thus has the same origin as the social malaise we are suffering from. It too is due to the lack of secondary organs intercalated between the State and the rest of the society. We have already seen that these organs seem necessary to prevent the State from tyrannizing over individuals: it is now plain that they are equally essential to prevent individuals from absorbing the State. They liberate the two confronted forces, whilst linking them at the same time. We can see how serious this lack of internal organization is, which we have noted so often: this is because it involves in fact something of a profound loosening and an enervation, so to speak, of our whole social and political structure. The social forms that used to serve as a framework for individuals and a skeleton for the society, either no longer exist or are in course of being effaced, and no new forms are taking their place. So that nothing remains but the fluid mass of individuals. For the State itself has been re-absorbed by them. Only the administrative machine has kept its stability and goes on operating with the same automatic regularity.

It is true the situation has many parallels in history. When a society is forming or being revived it passes through a similar phase. Indeed, in the final analysis, it is from the actions and reactions through direct exchanges amongst individuals that the whole system of social and political organization has been evolved. Therefore when it occurs that a system is carried away by time without any other taking its place as it disintegrates, it is inevitable that social life must go back in some degree to its primary source, that is, to the individuals, to be elaborated afresh. Since they stand alone, it is through them direct that the society has to operate. It is they who dispense the functions that once pertained to the organs that no longer exist or that will form part of those that are still needed. They themselves have to make good the organization that is lacking. That is our situation at present, and although there is a remedy, and one may see in it a necessary phase of evolution, we cannot disguise the critical element in it. A society made of a substance so unstable is liable to disintegrate if it suffers the least shock.

There is nothing to protect it against things from without or within.

All these considerations have been necessary in order to explain in what spirit the various civic duties should be understood, put into practice and taught. First, we have the duty that commands us to respect the law, and the one that prescribes our taking part in the elaboration of the laws of the land through our vote, or in more general terms, taking part in public life.

It has been said that respect for the law, in a democracy, has derived from the fact that the law expressed the will of the citizens. We should submit to it because we have willed it to be the law. But how could this hold good for the minority? Yet it is this minority which has the greatest need to practise that duty. Moreover we have seen that, in fact, the number of those who, direct or indirect, have willed any given law, represents only an insignificant part of the country, at any time. But without stressing the reckoning, this way of justifying the respect due to the law is the least suited to instil it. How does the fact of having willed a certain law make it worthy of my own particular respect? What my will has done, my will can undo. Mutable as it is in its nature, it cannot serve as a foundation for anything stable. We are sometimes surprised that the reverence for legality should be so slightly rooted in our consciousness and that we are always so ready to abandon it. But how can there be any reverence for any legal fiat that can be replaced from one day to the next by a different fiat on a single decision of a certain number of individual wills? How can we respect a law which may cease to be the law as soon as it ceases to be willed to be so?

The true source of respect for the law lies in its clearly expressing the natural inter-relation of things; the individual, especially in a democracy, will respect it only in the degree to which he recognizes it as having this quality. It is not because we have made a certain law or because it has been willed by so many votes, that we submit to it; it is because it is a good one—that is, appropriate to the nature of the facts, because it is all it ought to be and because we have confidence in it. And this confidence depends equally on that inspired by the organs that have the task of preparing it. What matters, then, is the way in which the law is made, the competence of those whose function it is to make it and the nature of the particular agency that has to make this particular function work. Respect for the law depends on the quality of the legislators and the quality of the political system. Here, the particular advantage of a democracy is that, owing to the communication set up between those governing and the citizens, the latter are able to judge of the way in which those governing carry out their task, and knowing the facts more fully, are able to give or withhold their confidence. Nothing can be more mistaken than the idea that the democracy has a right to our respect only in so far as it is pledged to the drafting of laws.

There remains the second point—the duty of voting. We need not here consider what this duty might become in some vague future and in societies better organized than our own. Quite possibly it may lose its importance. It is possible there will be a time when the appointments necessary to control political organs may come about, as it were, automatically, by the pressure of public opinion, and without, properly speaking, any definite reference to the electorate.

At the present time, however, the situation is quite different. We have studied its abnormal aspects, and indeed it calls for duties of a particular kind for this very reason. It is on the mass of individuals that the whole weight of the society rests. It has no other support.

It is therefore legitimate for every citizen to some extent to turn into a statesman. We cannot insulate ourselves in our professional callings merely because public life, for the moment, no longer has any agents to make it work except the manifold individual energies. But the very reasons that make this task necessary give it definition, too. The task is due to a lawless condition that has to be, not subdued, but treated to bring it to an end. Instead of offering this absence of organization, wrongly called democracy, as an ideal, a limit should be set to that condition. Instead of clinging to a jealous preserving of these rights and privileges, a cure has to be applied to the evil that makes them inevitable for the time being. In other words, the primary duty is to work out something that can relieve us by degrees of a role for which the individual is not cast. To do this, our political action must be to establish these secondary organs which, as they take shape, will release the individual from the State and vice versa, and release the individual, too, from a task for which he is not fitted.

X

DUTIES IN GENERAL, INDEPENDENT OF ANY SOCIAL GROUPING—HOMICIDE

WE now come to a new sphere of morality. In the earlier lectures we studied the duties that men have towards one another because they belong to a certain definite social group, or because they are part of the same family or guild or State. But there are other duties independent of any particular grouping. I have to respect the life, the property, the honour of my fellow-creatures, even when they are not of my own family or my own country. This is the most general sphere in the whole of ethics, for it is independent of any local or ethnic conditions. It is also the noblest in concept. The duties we are going to review are considered by all civilized peoples as the primary and most compelling of all. The supremely immoral acts are murder and theft, and the immorality of these acts is in no way diminished if they are committed against those of another country. The field of domestic morals, professional ethics and civic morals is certainly not so grave a matter. Whoever fails in one of these duties seems to us as a rule less of an offender than the man who commits one of the other offences. This idea is so general and so deeply engraved in men's minds that crime, to the ordinary consciousness, consists in essence or almost solely in killing, wounding or stealing. When we form an image of a criminal, it always takes the shape of a man who makes an assault on the property or the person of another. All the work of the Italian school of criminology rests precisely on the postulate, seen as an axiom, that this is the whole of crime. Establishing the type of an offender consists, for instance, in establishing the type of the homicide or thief, with their respective variants.

In this connexion, morals of the present day are in complete contrast with those of ancient times. A real transposition or reversal in the hierarchic order of duties has come about, especially since the coming of Christianity. In the altogether primitive societies and even under the regime of the City State, the duties we are going to consider, instead of being at the highest point of all morals, were only on the threshold of ethics. Instead of being set above all others, these duties, or some of them, had a kind of optional character. The proof of the slight moral dignity attached to them is seen in the lack of any severe penalties for their infringement. Very often, in fact, no penalty followed. In Greece, even the murderer was punished only on the demand of the family, who might be satisfied with a monetary indemnity. In Rome and in Judea, any coming to terms or compromise on homicide was prohibited, since it was considered a public felony, but it was not the same in the case of wounding or theft. It was for the injured individuals to pursue their own redress, and they could if they liked allow the guilty man to buy himself off by a sum of money. On the whole, therefore, there were only quasi-civic sanctions for such acts; very often they only entailed some sort of damages; in any case, even when the sentence on the guilty one carried a physical penalty, such crimes did not appear grave enough for the State itself to take repressive action. It was the individuals who had to take the initiative in doing this. The society did not itself feel immediately involved or threatened by these outrages that are repellent to us. It might even happen that this slight measure of protection was only given by the society to its own members and withheld when the victim belonged to another country.

The true crimes, then, were those carried out against the family or religious and political orders. All that threatened the political structure of the society, any shortcoming towards the public divinities, which were only the symbolic forms of the State, and any breach of family duties, were fraught with penalties that might indeed be terrible.

This development, that has had the effect of raising to the highest point in morals something that began by lying at the lowest, is a result of the parallel development that came about in collective sensibility, which we have often noted. Originally, the strongest collective sentiments, those that least tolerate opposition, are those concerned with the group itself, whether it be the political group as a whole or the family group. Hence the exceptional authority of religious sentiments and the severe penalties that ensured respect for them, the sacred things being but the symbol of the collective entity. This entity was personified in the form of God and of every kind of sacred being, and it is this collective entity that is the object of the respect and worship which appear to be offered to the imaginary beings of religion. On the other hand, social sensibility is not at all acutely affected by all that concerns the individual. The suffering of the individual makes little impact on the feelings, for his well-being is only of small account. Nowadays, on the contrary, it is individual suffering that is the hateful thing. The notion that a man suffers without deserving it is intolerable to us, but as we see, even suffering deserved oppresses and pains us and we try to ease it. The reason is that these sentiments that centre on man, the human being, become very strong, whilst those that link us direct with the group pass into the background. The group no longer seems to have value in itself and for itself: it is only a means of fulfilling and developing human nature to the point demanded by the current ideals. It is the supreme aim, compared with which all others are but of secondary value. That is why morals of individual man have come to transcend all others. We have so often pointed out the reasons underlying this decline of certain collective sentiments and the advance of others, that there is no need to review them again. They derive from those causes put together which, by increasing the diversities of the members of all societies, have left them with no essential characteristics in common except those they get from their intrinsic quality of human nature. It is this quality that quite naturally becomes the supreme object of collective sensibility.

The general features of the branch of ethics we are now tackling have thus been set out. We must next look at the detail in order to study the main rules, that is, the main duties it lays down.

The first and most imperative is the one forbidding any attempt on the life of a human being and prohibiting homicide, except in cases allowed by law (i.e. in case of war, in legally pronounced conviction and in legitimate self-defence). The reasons for the prohibition of homicide and for the increasing severity of those terms need not be dealt with here, after all we have said. Given that the goal of the individual is moral well-being and that to do good is to do good to others, it is clear that an act that results in depriving a human being of his life, that is, of the condition on which every other boon depends, must of necessity seem to be the most heinous of all crimes. We must not linger to give an account of the origin of the rule prohibiting murder. It is more to the point to trace how the rule operates in our present-day societies and the causes of the sway, great or small, that it holds over the consciousness, and of the respect, great or small, attached to it. In this quest, statistics must be brought in to help us. They will enlighten us on the conditions that cause the rate of homicide in society to vary: this rate indicates the degree of authority with which the rule prohibiting murder is clothed. This enquiry will make us understand the nature of the crime better and the same time throw light on the distinctive features of our morality.

On these considerations, it might seem true to say that the causes on which the tendency to homicide depends are obvious and have no need of being further defined. The reason why homicide is prohibited nowadays under threat of the most severe penalties provided by our code of law, is that the human person is the object of a sacred respect that was formerly attached to very different things. Should we not conclude that the reason why a certain nation has a bent towards murder in a greater or lesser degree is that this respect is widespread to a greater or lesser extent, and that a greater or lesser value is attached to all that concerns the individual? One fact seems to confirm this interpretation: ever since we have been able to follow the course of homicide by figures, we can see that it has gradually declined.

In France the rate for the period 1826–30 was 279; the rate decreases by degrees as follows: 282 (1831–35); 189 (1836–40); 196 (1841–45); 240 (1846–50); 171 (1851–55); 119 (1856–60); 121 (1861–65); 136 (1866–70); 190 (1871–75); 160 (1876–80); that is, a decrease of 62 per cent. in 55 years, a fall that is all the more remarkable since the population increased in the same period by more than one-fifth. We find the same decline in all civilized nations, to a greater or less degree according to the country. Thus it appears that with the progress of civilization homicide decreases. This seems to be confirmed by the other fact that it is the more widespread where countries are less civilized. Italy, Hungary and Spain are in the lead, followed by Austria. The three first-named are certainly amongst those that are least advanced: they are the backward countries of Europe. They stand in contrast with the nations that have a high standard of culture—Germany, England, France and Belgium, where homicidal crimes amount to between 10 and 20 per thousand inhabitants, whilst Hungary and Italy have a rate of more than 100, or say, 5 to 10 times as many. Finally, there is a similar varying incidence to be found within each separate country. Homicide is mainly a crime of rural districts. It is amongst workers on the land that we find the highest incidence. There is no doubt at all that the respect and the value attached to the person by public opinion grow with civilization. Might we then not say that the homicide rate varies according to the relative position of the individual in the mounting scale of moral ends?

We can be confident that this reading is on fairly sure ground although it be far too general. There is no doubt the growth of individualism has some connection with the decrease in the rate of homicide, but it does not bring it about direct. If it had this effect, it would be equally manifest in other kinds of criminal attack to which the individual is exposed. Thefts, fraud and embezzlement inflict pain on the victims that is sometimes as acute as purely physical injuries. A commercial fraud, a serious swindle, can often work more distress at one stroke in the evil effects they cause than an isolated murder. All these particular evils, instead of diminishing, only go on increasing as civilization grows. Thefts, which amounted to 10,000 in 1829, stood already at 21,000 by 1844, at 30,000 in 1853, at 41,522 in 1876-80, that is to say, an increase of 400 per cent. Bankruptcies rose from 129 to 971. There are also the physical offences, that show the same rise: first, sexual offences against children, and also assault and battery, which went up from 7—8,000 in the period 1829-33 to from 15—17,000 in 1863-69. Respect for the person should nevertheless protect it equally against injury by wounding and a mortal assault. That this increase could have come about, on the other hand, argues that this sentiment in itself can have had only a fairly weak inhibitive force. It cannot, then, be this sentiment, we might say,

that accounts for the inhibitive spirit that the homicidal
current encounters at a given moment. Amongst the circumstances that go with the advance of moral individualism, there
must be some that are especially inimical to murder but do not
reveal an equal antagonism to other forms of assault on the
person. What may these circumstances be?

We have seen that, parallel with the progress of the collective
sentiments whose interest is mankind in general, the human
ideal and the material and moral wellbeing of the individual,
there was a retreat, a weakening of those collective sentiments
whose interest is the group or family or the State (independently
of the advantage that individuals may derive from them).
These two movements are not only parallel but are closely
bound up. If the sentiments we attach to the individual in
general are growing, it is precisely because the others are
weakening and because the groups can no longer have any
purpose other than the interests of the human person. If
homicide is diminishing, it is rather that the mystic cult of the
State is losing ground than that the cult of the human being is
gaining. Indeed the sentiments that lie at the base of the cult
of the State are in themselves stimulants to murder. Furthermore, they are of great intensity, like all collective sentiments;
therefore when offence is given, they tend to react with a
force in ratio to their intensity. If the offence is a serious one,
it may lead the man touched closely by the offence to destroy
his adversary. It may be all the more likely to have this result
since these kinds of feelings, on account of their own particular
nature, are especially liable to silence all feelings of pity and
sympathy which in other circumstances would be enough to
restrain the hand of the murderer. For when these other
feelings are strong, those of pity are weak. When the fame and
greatness of the State appear as the supreme good, when the
society is the sacred and living thing, to which all else is subordinate, its importance so far transcends the individual that
the sympathy and compassion he may inspire fail to counter-
balance and curb the exigent demands of the offended sentiments that are so much more impelling.

When it is a matter of defending a father or of avenging a
God, can the life of a man count in the scale? It counts indeed
very little when offset against objects of such value and weight.

This is why political beliefs, the sentiment of family honour, the sentiment of caste, and religious faith—all these may often in themselves carry the seeds of homicide. The great number of murders in Corsica are due to the surviving practice of the *vendetta:* the vendetta itself, however, derives from the fact that the point of family honour is still a very live issue—that is, the feelings that link the Corsican to his clan still have great force. The reputation of the family name still stands above all else.

Not only may these various sentiments lead to murder, but they produce, when they are strong, a kind of chronic moral disposition, of itself in a general way inclining to homicide. Once, under the influence of all these moral states of mind, we come to attach such little value to individual life, then we become used to the notion that it should and can be sacrificed to all sorts of things. Then, a very slight impetus is enough to lead to murder. All these tendencies have in themselves something violent and destructive, that incline the individual in a general way to destroy: they predispose him to violent demonstrations and bloody acts. That is the cause of the uncouth harsh temperament that is a feature of the lower societies. It has often been held that this uncouthness is a remaining vestige of brutishness, a survival of the tooth-and-claw instincts of animal nature. In reality it is the product of a well-defined moral culture. Animals themselves are not as a rule violent by nature, but only when the conditions of their life make it necessary.

Why should it be otherwise with man? He has remained for long ages harsh to his fellow creatures. Not because he was close to the animal: it was the nature of the social life he led that so shaped him. The practice of pursuing moral aims that were foreign to human interests made him relatively insensitive to human suffering. All these sentiments just discussed can in fact only be satisfied by imposing suffering on the individual. The gods we worship live only on the privations and sacrifices to which mortals subject themselves. Sometimes, human victims even are exacted and it is this toll that expresses in a mystic form what society exacts from its members. We can imagine that such training over generations is likely to leave in the consciousness a disposition to cause suffering. Moreover, all these sentiments are, too, very vivid passions, since they will tolerate no opposition and brook no question. Characters formed in this way are therefore in essence a product of the passions: they are driven by impulse. Passion leads to violence and tends to break all that hampers it or stands in the way.

The decline in the rate of homicide at the present day has not come about because respect for the human person acts as a brake on the motives for homicide or on the stimulants to murder, but because these motives and these stimulants grow fewer in number and have less intensity. These stimulants are the very collective sentiments that bind us to objects which are alien to humanity and the individual, that is, which bind us to groups or to things that are a symbol of these groups. At the same time, I do not mean to say that the sentiments which formerly lay at the base of moral consciousness are destined to pass away; they will persist and must persist but they will be fewer and have far less strength than they had formerly. And this is what causes the rate of mortality by homicide to have a downward trend in civilized countries.

Moreover, this interpretation can easily be verified. If it is accurate, all the causes that reinforce these kinds of sentiment must increase the rate of murder. War is obviously one of these causes. It reduces societies, even the most cultivated, to a moral condition that recalls that of the lower societies. The individual is obscured; he ceases to count; it is the mass which becomes the supreme social factor; a rigid authoritarian discipline is imposed on all volitions. Love of country and the attachment to the group cast into the background all feelings of sympathy for the individual. What is the result?

Whilst theft, fraud and embezzlement decline appreciably, owing to various causes, homicide either increases or, at best, remains stationary. In France, in 1870, thefts decreased by 33 per cent., falling from 31,000 to 20,000 and robbery with violence from 1,059 to 871. There was only a slight fall in murder; the figure fell from 339 to 307. And yet this decline is only apparent and conceals a rise that is very likely appreciable. Indeed this decrease in general criminality in time of war, derives (in one respect, not to be over-rated or denied)—especially if the country is invaded—from a cause that must inevitably affect homicide; that is, the state of disorder in the administration of the law. The prosecution of crime must be less efficiently carried out when the country is invaded and all is upside down. That is not all. The age at which most crimes of homicide are committed is from 20 to 30 years. The incidence of homicide in this age group is in the ratio of 40 per million per annum. All young men of this age at that time had been called to the colours; the crimes they did commit, or would have committed in times of peace are not given in the statistics. If the rate of homicide fell a little in spite of these two causes, we can be certain that in reality it had seriously increased. The proof of this is that in 1871, when the troops were disbanded and the courts of law could attend to their duties in a more regular way (but without there being any great difference in the moral state of the country) a considerable rise is to be noted. After standing at 339 in 1869 and at 307 in 1870, the rate of homicide rises to 447, that is, an increase of 45 per cent. It had not reached such a high level since 1851, an exceptional year, as we shall see.

Political crises have the same effect. In 1876, the elections for the Senate and the Chamber of Deputies took place; the cases of homicide went up from 409 to 422; but in 1877, political disturbances becoming intensified—during the 'Sixteenth of May period'—a formidable increase took place. The rate rose at one stroke to 503, a figure not recorded since 1839. During the years of restiveness that run from 1849 to the time when the Second Empire was finally consolidated, the same phenomenon occurs. In 1848, the numbers for homicide stand at 432, at 496 in 1849, at 485 in 1850 and 496 in 1851; in 1852 the fall began, although the figures still remain very high until 1854. During the first years of the reign of Louis-Philippe, rivalries were violent amongst the political parties. Meanwhile the curve continued to rise from 462 in 1831 to 486 in 1832. The highest point in the century was reached in 1839 (569.)

It is a well-known fact that the Protestant religion is more individualistic than the Catholic. Every member can adapt his own faith more freely and with more reliance on himself or his own views. The result is that the collective sentiments common to all members of the Protestant Church are fewer and less strong, or, at least, that they inevitably take the individual as their chief concern. Now, the tendency to homicide remains without comparison stronger in Catholic than in Protestant countries. On an average, the Catholic countries of Europe account for 32 cases of homicide per thousand, the Protestant countries not even four. The three countries standing in the lead for the whole of Europe are not only Catholic, but dyed-in-the-wool Catholic—Italy, Spain and Hungary.

To sum up, then, a fruitful soil for the growth of homicide is a state of public consciousness arising from the passions—a state that has a natural echo in the consciousness of individuals. It is a crime made up of a lack of reflection, of instinctive fears and of impulse. In one sense, all passions lead to violence and all kinds of violence to homicidal forces, although these latter, especially, have an effect which goes beyond the individual in its ends. Therefore, the rate of homicide is the greatest proof that our immorality is becoming something less passive and more thought out, more calculated. Such indeed are the features of our immorality, which is remarkable for its astuteness rather than violence. These features of our immorality are at the same time the features of our morals. They, too, are becoming increasingly cold, reflective, rational; sensibility has a more and more limited part to play, and this is what Kant interpreted in setting passion beyond any morals. The moral act seems to us to-day to be an act of reason. Moreover there is nothing surprising in the alikeness we observe in the features of morals and of immorality. Indeed, we know that they are facts of the same nature and that they throw light one on the other. Immorality is not the opposite of morality any more than sickness is the opposite of health, both being different forms of one and the same state—two forms of moral life, two forms of physical life.

Thus, all that raises the temperature of the passions, in public life, raises the level of homicide. Public holidays and days of festival of course have the effect of intensifying collective life and of causing over-excitement. Out of 40 crimes of homicide noted by Marro, 19 were committed on public holidays, 14 on ordinary days, 7 were doubtful. It is a very limited number of cases to go on, but the predominance of public holidays is so marked that it could not be accidental. In the whole year, there are in fact only about 60 days of public holiday or festival. They should therefore account for one-sixth of the cases on other days of the week. The incidence in respect of public holidays is appreciably higher than for other days: on this random calculation of homicide, it follows that the rate of such crime must in general be very high indeed. A strict analysis of homicide suggests a similar conclusion. It is all the more surprising to find homicide thus associated with a particular state of activity, since so high a level of activity could ordinarily pass as normal. But this is exactly what follows from the fact that crime is not outside the normal conditions of life. By the very fact that a certain amount of activity deriving from the passions is always inevitable, crimes are always occurring. The main thing is that the rate should correspond to the existing state or state of mind of the society. A society without homicide is no more innocent than a society without passions.[1]

---

[1] The chapter ends with four illegible lines: the sense, however, seems complete without them.

XI

THE RULE PROHIBITING ATTACKS ON PROPERTY

WE shall now pass on to the second rule of morals of individual man; this rule protects not the life but the property of the human person, whatever his social group, against unlawful attack. The first problem we have to tackle is to know the causes that decided the laying down of this rule. How are we to explain this respect that the property of others inspires—a respect that the law endorses by means of penal sanctions? How does it happen that things should attach so closely to the person that they share his inviolability? It would demand long research and method to deal with such a question, which is simply the origin of the right of ownership. But some outstanding points at least can be fixed.

Let us begin by studying the most usual solutions. The problem is to know what the bond consists of that links objects exterior to him with the person, objects which of course do not form part of himself. How does it happen that a man may dispose of certain things as he may dispose of his body—that is, to the exclusion of all else, since it is the lawfulness of this exclusion which makes the encroachments of others unlawful. The simplest and most radical answer would be that this bond could be resolved into elements and this means there was some element in the nature of man, some constitutional peculiarity which logically implied the allocation and appropriation of certain things. The notion of property could be deduced from the very notion of human activity. We have only to analyse this activity to discover why man is and must be an owner of property. The theory has been held by many thinkers that the idea of labour meets the case. In fact, work is the work of man: it displays the abilities of the individual and is no more than the person in action. Therefore it has a right to the same sentiments as those inspired by the person. On the other hand, it tends by its nature to be externalized, to be projected beyond itself, to be embodied in exterior objects whose whole value it creates. Here then we have things which are just human activity in crystallized form. There is no need, then, to wonder how they become attached to the person whose product they are, since they derive from and form part of, himself. He owns them just as he owns himself. Here we have not two different and heterogeneous terms, between which a third has to be found, to unite them; there would be a complete sequence between the two, and the one would be only a particular aspect of the other. "Nothing is implied in property" says Stuart Mill "but the right of each to his (or her) own faculties, to what he can produce by them, and to whatever he can get for them in a fair market." (*Polit. Econ.*, Vol. I, 7th Ed., p. 270.)

The postulate on which this theory rests seems so self-evident that we can find it at the basis of the most varying systems: the socialists invoke it just as the classical economists do. Yet it is far from being a truism. Let us take the proposition in itself, without considering the conclusions to be drawn or the application that is made of it. It is said that we should be free to dispose of the fruits of our labour because we are free to dispose of the talents and energies involved in this labour. But are we able to dispose of our abilities with such freedom? Nothing can be more debatable. We do not belong to ourselves entirely: we owe something of ourselves to others, to the various groups we form part of. We give them and are bound to give them the best of ourselves; why should we not be equally bound (and with even better reason) to give them the material products of our labours? Society takes years of our life and on occasion, requires our life of us. Why should it not be justified in requiring of us these external belongings of ours? The cult of the human person in no wise excludes the possibility of such obligations. For the human person whom this cult concerns is the human person in general; and if, to realize this ideal, it were necessary for the individual to yield up in part or entirely, the creations of the work which he had toiled over, this act would be a strict duty. Thus, in order that property be justified, it is not enough to invoke the rights that man has over himself; these rights are not absolute but limited by the claims of the moral aims, in which a man has to co-operate. It must therefore be made quite clear that these claims demand that the individual should dispose freely of the things he produces. In any case, there are many circumstances in which a man's right to free disposal is withdrawn: that is, when he is not able to make effective use of it, when he is not of age, when he is insane or when he is a confirmed spendthrift. This means that the right is subject to circumstances and is not a matter of course.

Let us go further, and accept this postulate. In order to justify property it should, to begin with, be quite differently framed. Property, in fact, is not exclusively acquired by labour, but may be derived from other sources :

(1) by exchange, (2) by donations *inter vivos* or legacy by will, (3) by inheritance.

Exchange is not labour. True, if it is done with strict equity, an exchange does not enrich, since the values exchanged are supposed equal. If they have been entirely created by labour, nothing will have been added to the property of those exchanging; on both sides, all they possess has been made by labour, either direct or indirect. But for this to be so, the exchange must have been absolutely equitable and the things exchanged balance exactly. To achieve this, many conditions would have to be fulfilled that do not exist in societies of the present day. It is even doubtful if they could ever be entirely realized. In any case, here we have property subject to a condition other than labour, that is, to the equity of contract. This simple interpretation of Mill does not, then, suffice in itself. In the second place, even if the system of contracts were changed in a way to satisfy all demands of absolute justice, property could still be acquired by other means which can in no way be related to labour.

Take inheritance, to begin with. The inheritor is endowed with goods and chattels of which he is not the originator and which he does not even owe to any act of the one who did create them. In certain circumstances, it is kinship alone that confers the right to property. Should we say that inheritance, no matter how it is governed, is a survival of the past that should disappear from the statute-book? There still remain donations and legacies by will or otherwise. Mill, who admits that inheritance goes against the moral notion of property, believes on the other hand that the right to bequeath by will or dispose of property by favour, is logically implicit in the notion. "The right of property", he says, "implies the right to give at will the product of his labour to another individual and the right of that other to receive and enjoy it." (*ibid.* p. 273.) If property, however, is to be respected and normal only where it is based on labour, how could it be justified when it is based on a legacy or donation? And if it be immoral to acquire it by way of a gift by favour, the practice in general of gifts would be condemned by that very fact. But, it is said, does not the right to possess logically carry with it the right to give? This proposition is not in the least self-evident; the right to enjoy the things we possess has never been absolute; it is always hedged about with restrictions. Why should one of these restrictions not have a bearing on the right to give? In fact, the right to give is already limited. An individual is not allowed to dispose of his property by fixing in advance to whom it should go after the death of the beneficiary. The right of gift or donation can therefore only be exercised to the benefit of one generation. This means that there is no immunity about the right. So there is no inherent anomaly in its being even more strictly limited. Furthermore, Mill himself admits that a limitation is needed, precisely because it is neither moral nor socially desirable for men to get richer in this way without any effort of their own. He proposes to decide the quota of what anyone might receive by way of a legacy. "I see nothing objectionable in fixing a limit to what any one may acquire by the mere favour of others, without any exercise of his faculties." (*ibid.* p. 281). This is an admission that a gift or donation contravenes the principle whereby property is the product of labour.

Thus, if we admit this principle, we have to say that property, as it exists at present and as it has been since the beginning of societies, cannot in the main be justified as an institution. True, it is extremely likely that property will not be in the future what it has been up to the present; but the right to say that this or that form of it must disappear, requires more from us than merely showing that these forms conflict with an earlier principle. There still remains to demonstrate how they were able to establish themselves and under the influence of what causes, and to prove that these same causes are no longer actually present and active. We cannot demand that existing practices be put down on the score of an *a priori* axiom. Is it indeed possible that any contract could be absolutely equitable, or that there could be any society in which all gifts or donations would be prohibited? These are big problems, for which it is difficult to presume any answer. In any case, before knowing what property ought to be or should become, we should first have to know how it has become what it is and how we can account for the form it has in present-day societies. To this question, the theory of labour gives no answer at all.

But to go further, I would say that in itself labour could in no case become the primary cause of property. In every age, we see that labour does not produce the raw materials to which it is applied and that it presupposes implements or, at very least, physical media that it has not produced. The answer to this objection has been that these physical media have no value in themselves; that they must have been previously elaborated by the hand of man. We must recognize, however, that if these media are to acquire their value, they need to be elaborated to a greater or lesser degree, according to their state of development, and this in turn calls for labour of greater or lesser intensity. To get all it can yield out of the land, little effort is needed when the soil is fertile, but a great deal when it is not. And so properties that are equal in value may be founded on measures of labour that are very far from being equal. This means that in the one instance, the labour is replaced by something else. Furthermore, even when the natural means are in themselves without value, labour divorced from these means must needs be fruitless. So that the labour presupposes something other than itself, some point to which it must be applied, and that is this virtual value which the labour translates into the act. And this virtual value is not a negative one. The objection mentioned above might however apply in a more general way. In reducing property to terms of labour, we admit that the value of things derives from objective and impersonal causes, not subject to any appraisal. But nothing of the kind. The value depends on opinion and is a matter of opinion. If I build a house in a spot that suddenly becomes sought after for its qualities of charm or some other reason, its value becomes much enhanced. If, on the contrary, favour turns against it, it may reach the point of having no value at all. A mere whim of fashion can make this or that object or stuff, for instance, and hence the natural means employed in manufacturing them, rise in price. My property, or what I own, might double its market value without my lifting a finger. On the other hand, the moment the new improved machines were invented, Jacquard machines became thereby valueless. The man who owned them was in the same position as one who owned nothing, even if he had had them built entirely by the fruits of his own labour. Thus, in all property, there is an element over and above the labour of the owner, even if the object owned is the product of his own hands. Besides the initial contribution—that of the raw material, there is an element in it that comes from the society. According to the trend of social tastes and needs, so our property may grow or decrease as we hold it, although we have no hand in these fluctuations. Shall we say that it is necessary and even imperative that things should be so, and that these positive or negative shifts must be, if the society is to be well served? Must there not be a stimulus to individual initiative and the spirit of novelty, and a kind of penalty for the spirit of routine and inertia? It is certain that however economic life be organized, the value of things will depend always on public opinion and it is a good thing this should be so. Nevertheless, the fact remains that there are values, too, and hence, elements in property that do not derive from labour. We may even admit that sometimes these elements are the fruits of some timely foresight, and we can see in them the expression of a natural talent. How often it happens that these elements become an addition to the things we own, or are taken away from them, owing to some simple encounter or by pure chance. I could not foresee that a main highway was to run alongside my property, but this brings an increase in value, that is, it grows automatically. And again, drastic changes in machinery may bring ruin on the real property of a manufacturer.

So it is in vain that we may try to make any deduction about things, starting from the person. These terms are heterogeneous. The law that associates them is synthetic in its effect. There is a cause exterior to them, which brings about their association.

Kant was well aware of this fact. It is true, he says, that if we see in property only a material holding, it is easy to demolish it by a process of analysis. If I am linked physically to a certain object, if I keep it in my hands, for example, any one who lays hold of it without my consent is trespassing on my internal freedom. "The proposition expressing the principle of an empirical rightful possession does not thus go beyond the right of a person in reference to himself." (*Princ. of Priv. Right*, para. vi.) But how does it happen that I can call a thing my own when I do not possess it physically? In the first case the thing made one with myself; now, it is separate, it is something different from myself. It can, then, only be attached to me by a synthetic bond. What forms the basis of this bond? (*ibid.* para. vi.)

Such a bond, by definition, can only be an intellectual one, since it is independent of any condition imposed by time or place. Since the thing remains independent of my person whatever the place I may be living in, this dependence must have its source in some mental state which is itself, in some way, beyond space. I may say that I own a field—one that is situated where I myself am not: "the question here is only one of an intellectual relation between the object and myself." This relation is founded on an act of my will. Indeed, my will also is free of any spatial condition; what the will decrees is valid and binding for all men alike, wherever they are placed. The will determines their relations, independent of any place they may be in, for it is universal. It is beyond what is perceptible by the senses and therefore the rules it lays down do not allow of their application being limited by any condition that is perceptible by the senses. This proposition is especially obvious if we admit the principles of the *Critique*. Indeed, according to Kant, whilst intelligence and thought are subject to the laws of time and place, it is quite otherwise with the will.

Thought is related to phenomena; it is in the world of phenomena, and the mind cannot apprehend phenomena outside of a spatial or temporal *milieu*. But the will is the faculty of the noumenon, of the entity or thing in itself. The will, then, is outside these phenomenal appearances, to which, therefore, its actions cannot be subject. If, then, I am justified in willing an object exterior to me to be mine, this act of my will is valid whenever I happen to be in space, since space is unknown to my will. Since, on the other hand, my will has a right to respect, whenever it is legitimately exercised, it is enough that my will shall have legitimately decided to declare this object its own, for the appropriation in itself to be valid *de jure* and not only *de facto*. So it would be this particular feature (by virtue of which my will is to be respected and to have sanctity for others whenever it is exercised without violating the rule of law) which alone could create an intellectual bond between these things and my person. It is clear, however, that this interpretation can be dissociated from the hypothesis of the *Critique* and maintained by some other method. For, in general, it is recognized that a decision of my will is not subject to laws, like the movements of my body. By my will, I can free myself of space. I can will a thing to be mine independently of any local situation. The essential point of this doctrine is not the philosophic theory on the relativity of time and space: it is the idea that when my will has asserted itself according to its right it must be respected; in a word, it is the sacred character of the will, so long as it itself conforms to the law of conduct.

But, as we see, the interpretation is not complete. It still remains to demonstrate that I may will an object to be my own without failing in the principle of the right, whereby this exercise of my will is legitimate. Kant invoked another principle to make this clear.

Let us first clearly define the range of the right that I thus assume as mine.

"If, by word or deed, I declare my will that some external thing shall be mine, I make a declaration that every other person is obliged to abstain from the use of this object upon which my will is exercised. . . . But this claim assumes that I, reciprocally, acknowledge myself bound to observe a similar abstention towards every other in respect of what is externally theirs. I am therefore not bound to respect what another person declares to be his, unless every other person likewise guarantees for his part that he will act in relation to what is mine on the same principle." (*ibid.* para. viii.)

My own individual will, being only individual, cannot lay down the law for others. Such an obligation cannot be decreed except by a collective will set above each individual will taken separately. "I cannot, in the name of my own individual will, oblige any one to abstain from the use of a thing, in respect of which he would otherwise be under no obligation; I can do this therefore only in the name of the collective will of all united in a relation of common possession." (*ibid.* para. viii.) Each one must be bound by all, and a collectivity can bind its members concerning a given thing only if it has rights over this thing, that is, if it possesses it, collectively. We thus arrive at the following conclusion: in order that men may be justified in wishing to appropriate individual things, those things must have been originally possessed by a collectivity. And since the sole natural collectivity is that made up of the whole human race (since it is the only one that is complete, all others being only partial), the right of the intended appropriation implies an original community of things and derives from this fact. If we leave aside the idea of this common possession, the binding and reciprocal character presented by individual ownership becomes unintelligible. To what degree and in what sense has this original community of things any logical basis?

Let us suppose that the earth were an infinite plain, and that human beings were dispersed over it in such a way that they did not form any community amongst themselves: in these conditions it would not be possible for them to have any possession in common. But the earth is spherical and hence of limited surface area. Men are thus compelled by the unity of habitat to be in relation—thus they form a whole and this whole is the natural owner of the total habitat which it occupies, that is, of the earth. "All men are originally in rightful possession of the soil. . . . This possession is a possession in common, because of the connexion with each other of all places on the surface of the earth as a globe.' (*ibid.* para. xiii.) Thus, in the beginning, the sole rightful possessor was mankind. In what way can mankind logically exercise this right? There are two different ways of interpreting this, and two only. Either mankind may declare that, since all belongs to it of right nothing could belong to any one person. Which is absurd; for if individuals do not exercise the right of collective ownership, it might as well not exist. This would in practice amount to denying the right of collective ownership. Or, mankind may acknowledge the right of each one to appropriate all that he can, with the reservation of concurrent rights of others. On this basis, the right would become a reality and pass to the act. Thus, the originating community of the soil was able to come about only by way of the appropriation we are discussing, with the reservation just mentioned. Herein lies the basis of our right to 'will to have' an exterior object 'as our own'.

One last condition remains to be settled. I cannot, by virtue of the right I derive from mankind, appropriate any thing except on condition of not trespassing on the similar right possessed by others. How can this condition be carried into effect? It suffices and it must be that my appropriation shall be antecedent to that of others; that it should have the advantage of priority in time. "The act of taking possession . . . can accord with the law of the antecedent freedom of any one only on condition of having the advantage of priority in time . . . that is, of being the the first act of taking possession." (*ibid.* para. xiv). Once my will is declared, no other may declare himself in a contrary sense; but on the other hand, if no other will has been declared, mine may assert itself in complete freedom. Since my will to appropriate is asserted by occupancy, the condition on which the legitimacy of my appropriation depends, is to be the first occupier. With this reservation, no limit is set to my right. I may extend my possession as far as my powers allow. "The question arises of knowing how far the right goes, of taking possession of the land; I would reply: as far as the ability of having it in his power, that is, as far as the one who wills to appropriate it can defend it. It is as if the land were to say: 'if you cannot protect me, neither can you command me'." (*Ibid.* para. xv.)

Let us sum up. The human race is the ideal power of the soil. This right of ownership can only become reality through individuals. On the one hand, individuals have the right to will to appropriate all that they can of the common demesne, with the reservation that none shall trespass on the rights of others, a condition fulfilled by the single fact that the land appropriated is so far unoccupied. On the other hand, because the act by which this appropriation is made is an act of will, it is free of any relation to space. Thus it has the same moral value, whatever the place in which object and subject respectively may be situated. In this way, the possession of some thing which I do not actually hold becomes justified. At the same time, we must add that this justification holds good not only *de jure* and in concept, but *de facto*. What lends support to this argument is that mankind ought logically to have the will for individuals to appropriate things in this way; but this entirely logical and ideal right only involves corresponding obligations in turn. It authorises the individual to repel any unlawful trespass, but it does not place at his service any ready means to enforce this right, for mankind is a group only in concept, just as it is proprietary only in concept. The alternative is that there must be true groups, or groups *de facto*, established, to protect the rights of each one. In other words, there are, to use Kant's expression, no decisive acquisitions except within the Civil State of society (*ibid.* para. xv.) But this does not mean that it is only in the Civil State that there is a basis for the right of ownership: it only recognizes and guarantees it. These rights are founded in the nature of things, that is: (1) in the nature of the will, and (2) in the nature of mankind and man's relationship with the earth he inhabits.

What gives weight to this theory is that it carries justification for the right of the first occupier—the most precise ever attempted: this is done in the name of the principles of morals that are in the main spiritualistic. "Labour forms no more than an external sign of possession." (*ibid.* para. xv.) In short, if we strip away its dialectic, the theory can be reduced to quite simple terms. It is unreasonable and contrary to the character of mankind that things should not be taken possession of; any appropriation is lawful which is carried out on land that has already itself been thus appropriated; and the will that governs this action has the right to be respected once it has declared itself, even when the individual (or subject) and the thing are not in contact. We find here (as in Kantian morals as a whole), two principles associated and combined, although seemingly contradictory. One is the principle of the immunity of the individual will and the other, one that makes the individual will subordinate to a law superior to it. It is this superior law, in the end, that welds together the two heterogeneous entities that have to be united to form the notion of property. In that respect, this seems to us to have advantages over the theory concerning labour. It gives a better indication of the difficulty of the problem; it lays down clearly the dual nature of the two terms and states precisely where we get the third that acts as a link, that is to say, the collective will to which the separate wills are subject. The weakness of the theory lies in assuming that the priority of the occupancy is enough to give this occupancy a juridical and moral basis; that the wills do not mutually cancel each other out and do not mutually trespass on one another because they do not run against one another on the physical plane in respect of the same object. It is contrary to the principle of the theory to be satisfied with this exterior and physical compatibility. The wills are all that they are able to be, being independent of manifestations in space. If, then, I appropriate to myself an object which is not yet appropriated in fact, but which is already willed by others—and this will has been declared—I am denying the force of this will morally just as certainly as if material trespass had taken place.

XII

THE RIGHT OF PROPERTY (continued)

KANT's theory may be summarized in this way. The earth is the property of the human race. A property that is not appropriated is not a property. It would therefore be unreasonable and inconsistent that the human race should forbid the appropriation of the soil. That would be a denial of its own rights. But this appropriation can be made only by men, either individually or by little groups. Thus, the right that mankind has to the earth implies the right of individuals to occupy limited parts of the earth's surface. On the other hand, since the will, when its decisions are legitimate, has a right to respect, any first occupancy is to be respected and the consciousness of mankind has to recognize its legitimacy. For my will, in acting thus, is only making use of its right, without trespassing on any other right, since, by hypothesis, no other separate will had yet possessed itself of the same object. The right that I derive from mankind, that is, in short, as a human being, can be limited only by the similar right of other men. If, then, other men have not asserted their right concerning the things that I appropriate to myself, my right over them is absolute. Hence it follows that I have the right to appropriate to myself all that I can of the things which have not previously been the object of any appropriation. Within these limits, my right extends as far as my power. Since the dictates of my will derive their force from my will itself and this will is beyond the confines of space, the act by which I declare myself to be the owner of a thing makes me the owner, even if I do not hold it physically.

What gives this line of argument its weight is that we find in it a moral theory of the right of the first occupier. Kant does not at all shirk this consequence of his theory. He does not hesitate to claim as his own the well-known saying: 'fortunate, he who is in possession', *beati possidentes*. For this privilege however, usually put forward as a social necessity or convention or fiction, he undertakes to find a basis in law. "This prerogative of right that follows from the fact of empirical possession according to the formula of *beati possidentes*, does not come about because the possessor, presumably an honest man, is not bound to prove that his possession is legitimate but because . . . everyone is invested with the faculty of having as his own any object exterior to his will." (*Princ. of Priv. Right*, para. ix.) There is, then, an element in the concept of property here set out quite different from that of labour, and that is why it is important to appreciate this theory which, in itself, makes us feel what there is one-sided about the first theory. There can indeed be no doubt that in any occupancy not contrary to a pre-existing right, there is an act that confers certain rights. At all times in history, mankind has been wont to grant privileges of right to a first possession. The declaration of will by which we assert our intention to appropriate an object that has no actual owner at the time, is not without moral value and has a right to some respect.

On the other hand, the impossibility of reducing all property to this sole element is especially striking in a theory that attempts to base the prerogatives of the first occupier on a moral principle and not solely on utilitarian grounds. Kant is compelled to contradict his own reasoning. If the wills are exercised to the full, independently of any manifestations in space, they may find themselves in conflict without being physically engaged. They may cancel each other out, contradict one the other and mutually repel one another, when even the bodies that they impel do not clash at any given point in space. If I appropriate to myself an object which has not yet been appropriated in fact by someone else, but which he wills to be his, without this will having found physical expression, is there no case of trespass by one on another? For there are no objects which are not liable to be willed by people other than the one who in fact takes possession. Material obstacles or some physical impossibility may have prevented the other party from asserting himself in time and going ahead. It is impossible to grant any moral value to chance encounters or to a purely physical advantage. There is in fact something almost logically preposterous about such considerable space being given to material force in a spiritualistic theory. The range of things that I can legitimately appropriate is determined solely by the reach of my power. "No one shall within the range of a cannon-shot indulge in fishing or in collecting yellow amber along the shores of a country which already belongs to a certain State." (*ibid.* paras. xv–xvii.) Here we see the lawfulness of appropriation dependent on the range of a gun. Thus, should guns of longer range be introduced, the lawful juridical demesne of the State would be extended *ipso facto*.

Precisely because the act of volition is a mental act, the equilibrium of individual wills has itself to be a thing of the mind, that is, moral. It is not justified solely by the fact that the material movements by which these wills are expressed are some of them geographically exterior to the others and that they do not meet at the same point in space. They must first of all not cancel each other out or exclude each other morally. In order that the occupancy shall be considered legitimate, at least as far as our actual conscience is concerned, it has to be subject to conditions other than simple priority in time. We cannot acknowledge the individual to have the right to occupy all that he can physically occupy. This was already admitted by Rousseau. He too traced the right of property back to the right of first occupancy hallowed and sanctioned by society. He, however, limited the rights of the occupier to his normal needs. "Every man", he said, "has a natural right to what he needs." (*Contrat Social*). One man intrudes, therefore, on the right of another solely in appropriating more things than he has need of, even when these things are not yet appropriated. "As a rule", he says, "in order to give authority to the right of the first occupier over any piece of land whatever, there have to be the following conditions: first, that this land be not yet occupied by anyone; second, that a man occupy only the area required to subsist." Rousseau adds, it is true, that labour and cultivation are necessary for a genuine taking of possession. This does not mean that the labour seems to him to imply by analysis the right to possess, in accordance with the theory we discussed in the first instance. It means that in his view, labour is the sole authentic symbol of occupancy. It is no more than a symbol, a title in law. On this point, then, he only diverges to a slight degree from the theory of Kant.

A more important divergence occurs in his making the area of lawful occupancy subject to the area of normal needs. But in this instance, clearly, whilst the right of the first occupier in Kant's theory was not limited enough, here we find it too restricted. We might say, perhaps, that a man has a right to possess at least what he needs for existence, but not that he has not the right to possess more than that. Rousseau was dominated by the idea that there is a natural state of balance governed by the nature of man, on the one hand, and the nature of things, on the other; that any upset of this balance caused a decline in the normal state of man and threw him into misfortune and pain. Hence his conception of a society where all conditions would be perceptibly equal, that is, equally of an average, each one possessing not much above the bare essentials for existence. But to-day, this concept has no more than historical interest. The hypothesis of natural equilibrium has no reality. The great change that social life brings about is to put in the place of a stable unvarying equilibrium such as we see in animals, an unstable one that varies constantly; and further, to have put in the place of so-called natural needs, others that it is not imperative to satisfy in order to maintain life, but whose satisfaction is not any the less legitimate.

This discussion, at any rate, has had the advantage of letting us see the complexity of the phenomenon. It is certain that many and varying elements enter into it: let us try to analyse them. To do this, we must first define the thing we are speaking of. What should we understand by 'right of property'? What does it consist of? How is it recognized? We can see that a solution to the initial problem will make it easier to examine the causes.

The definition we are seeking has to express the right of property in a general sense, that is, allowing for the particular forms it may have taken in different ages and countries. We should first try to get at the essence of this right, or what is common to the various ways in which it has been conceived.

The idea of property first evokes the idea of a thing. There seems to be a close connexion between these two notions: that one can only possess things, and that all things can be possessed. Undoubtedly, in the present state of our ideas, we find it repugnant to admit that the right of ownership may be exercised over objects other than things. At the same time, we have to include animals within the term of things, since they may be as absolutely possessed as inanimate things. But this limitation is comparatively recent. As long as slavery existed, slaves were the subject of a 'real right' that cannot be distinguished from the right of property. The slave stood to his master in the same relation as his fields or his beasts. In some respects it was the same thing with the son of the family, at least in Rome. Except in the matter of his civic commitments, he was considered as an object of ownership or chattel. In ancient times he could be ransomed. Now, the *rei vindicatio* applied only to things that formed a citizen right of ownership, that is, to corporeal property *in commercio*. In the classical era, the father could still dispose of him validly as property and until the time of Justinian he could be the object of a *furtum* or theft. The notion of this disposal of property and of *furtum* must needs imply a thing subject to the right of property.

Conversely, there are things which are not the object of any kind of right of property. Such were formerly the sacred things, known in Rome as the *res sacrae* or *religiosae*. The sacred things were in fact outside any transactions, absolutely inalienable and could not become the object of any real right or any obligation whatever. They were not owned by anyone. It is true we might say, and it was said, that they were the property of the gods. But the effect of this very formula is that they did not constitute human property, and we are concerned here with the right of ownership exercised by men. This attributing of the sacred things to the gods was in reality only a way of declaring that they were not and could not be appropriated by any man. But that particular feature is not confined to this category alone. There was also what was called in Rome the *res communes*, that is, things which belong to no one because they belong to all and by their nature elude any appropriation: the air, springs and streams and the sea. Everyone may use them, but no individual or group can be pointed to as the owner. There exists to-day what is called property under public ownership—roads, highways, streets, the banks of rivers that are navigable or used for timber rafts and the shores of the sea. All these forms of property are administered by the State but they are not owned by it. Everyone makes free use of them, including foreigners. The State that manages them has no right to part with them; it has to carry out obligations connected with them but has no right of property over them.

What emerges from these facts is that the range of objects liable to appropriation is not necessarily settled by their natural composition but by the law of any nation. It is public opinion in every society that makes some objects regarded as liable to appropriation and others not: it is not their physical nature as natural science might define it, but the form their image takes in the public mind. A certain thing which yesterday could not be appropriated, may be so to-day and vice versa. Hence it follows that the nature of the entity appropriated cannot come within our definition. We cannot say that it should consist of something corporeal that is perceptible by the senses. There is no reason why incorporeal things may not admit of appropriation. *A priori*, no limits can be set to the power the collectivity has to endow anything that exists with the qualities requisite for juridical appropriation, or to take away those qualities. If I use the term 'thing' in what follows, it is in a sense that is strictly indeterminate and without wishing to prove the particular nature of the thing.

One may say the same of the subject or individual who possesses. True, it is usually a man or men who own property. But in the first place, it may be sometimes a group or perhaps a legal entity like the State, the *commune* or the family, in so far as the ownership is a collective one. So it is not only a man who has been able to own property. For long ages, only the individual member of every society could exercise this right. The range of persons qualified to own is decided by the laws of each country, as is the range of things qualified to be owned. Therefore all that we can define is the nature of the links that bind the thing appropriated to the subject or individual who appropriates it, and leaving out of account all that goes to make up the characteristics of both. Of what do these links consist, and what are their distinctive features?

At first glance it might seem that the most natural course would be to look for these features in the nature of the powers in the hands of the individual who possesses, in respect of the thing possessed.

By legal analysis, these powers have long since been reduced to three: the *jus utendi*, the *jus fruendi*, and the *jus abutendi*. The first is the right to make use of the thing as it is—to live in a house, to ride a horse, to walk in a forest, and so on. . . . The *jus fruendi* is the right to the yield of the thing, the produce of the trees or the soil, the interest on a sum of money owned, the rent of a house, etc. . . . As we can see, there is only a shade of difference between the two. Both consist in the power of making use of the thing without altering its nature materially or juridically, that is, without modifying its physical composition or its legal status. It is this latter power that is implied in the *jus abutendi*. By this we mean the faculty of transforming the thing and even of destroying it, whether it be by consuming it or otherwise, or, again, of disposing of it to another or of changing its legal status. Does the setting out of these various powers reveal a true picture of the right of property?

In the first place, the power to make use of the thing is so little characteristic of this right that it may be exercised over things that do not admit of appropriation. I use air, water, all that is common property, and yet I do not own them. In the same way, I make use of highways, streets, rivers and so on . . . without owning them. I may pick fruit from trees that overhang the roads or that grow in the State forests without their being my property. In other words, the right of usage of a thing or of what it yields implies only that the thing in question has not been previously appropriated by another, but does not assume in any way that I have appropriated it. But then, what right of usage are we discussing? Is this an unlimited right? Should we say that the owner can make use of a thing as he pleases, without any limit being set? There is no country in which such freedom from limitation has been recognized or given sanction by law. The right of usage is always clearly defined and limited. There are regulations covering many things, such as getting in the crops, to which the proprietor is obliged to conform. In earlier times, it was strictly prohibited to carry the grain crops or harvest the vineyards before the appointed day, and the way this had to be carried out was also laid down. The right of usage of his property and its yield is extremely limited and yet is indeed the right of that owner. The same thing applies to the married woman who is the owner of her dowry, in respect of the substance and what it yields.

We might say the same about the *jus abutendi*, that is the right of disposal by . . . or otherwise . . . (*omission*). It may be exercised by people other than the proprietor. Any powers of management imply a power of disposal. The family council[1] or legal guardian may alienate or transform the property of a minor or an incapacitated person; they are not the owner of the property; and this applies also to the powers of the husband concerning the property of his wife. Very often the right of ownership in no way implies the right to alienate. For centuries the family patrimony has been inalienable: so long as the right to bequeath by will was not admitted, the rights of the head of the family to alienate his possessions were limited: he could not dispose of them freely by will. Nowadays his rights in this respect are still limited in most countries. The donations *inter vivos* which he was able to make have even been revoked by a number of laws in the case of surviving legitimate children. Real estate included in a dowry cannot be disposed of or mortgaged by the husband or by the wife—(although she be the owner), except in certain cases fixed by law. And this inalienability is so much in the interest of the wife that it allows of certain deeds being revoked, once the marriage is dissolved. The rights of one who has come of age but has a legal guardian are equally limited in this respect. What is most striking is the severe limitation set to all proprietary rights accorded to an owner. This is most clearly seen in their not being left to his discretion: he cannot preserve the rights in full unless he makes use of them in a certain way. The spendthrift, that is, one who makes use of his means irresponsibly, who dissipates and jeopardizes them, has their management and even enjoyment taken from him.

Thus, the power of use and enjoyment is found also in cases where there is no right of ownership. The power of disposal may be almost complete, without the right of ownership being thereby extinguished and may be exercised by those other than the proprietor. So that a list of all these rights will not enable us to recognize the characteristics of the right of property. Some of these may be missing, some may be found elsewhere, and all may vary greatly, but the right of property does not vary to the same degree. All that can be said is that wherever this right exists, there is a subject (or individual) who is capable of legally

[1] In France : legally constituted, under a *juge de paix*.

exercising certain powers over the object that is said to be possessed, but without it being possible as a rule to say with accuracy what these powers consist of. They must always exist, but we cannot say exactly what they are. Here, we may find the right of alienation and there it may be lacking; here, the right to alter the nature of a thing, and elsewhere there is none; here the right is fully developed and there it is less so, etc. . . . Seen in these terms, then, they cannot serve our purpose in defining the right of property. For there are some things over which we have certain powers and which we nevertheless do not possess. A mortgage gives rights over real estate that is mortgaged, but we are still not its owners; any right of management implies a certain control exercised over things, and yet to manage is not to possess.

A clear specification of the powers implicit in property does not enable us to define it. These powers are either too precise, too specialized, or too general. Either they are peculiar to certain forms of ownership or to certain circumstances, or they exist outside of any kind of property. One feature, however, is distinctive. We can make use of things we do not possess and do so legitimately, but when it is a matter of property, it is the owner alone who can use it, whether he be an actual person or a legal entity, an individual or a collectivity. The powers of usage may be either wide or limited, but it is he alone who may exercise them. A thing over which I exercise the right of property is a thing which serves myself alone. It is a thing withdrawn from common use for its use by a given subject. I may not have unrestricted enjoyment of it but no one other than myself can enjoy it. A legal guardian may be appointed who will supervise and plan the way in which I make use of it, but no one can be substituted for myself, to use it in my place. Or again, if there are ten of us to make use of a thing, this means that there are ten owners. To this it will be objected that there is such a thing as usufruct. Any one having this life interest does indeed enjoy the thing but is not its owner. But what is the element that makes the owner be its owner or an owner if if it is not that he is destined to enjoy it one day? Withdraw this right of eventual enjoyment and you leave him with nothing. A man is said to own landed property; this means that he is legally entitled to make use of the estate from a specified date. To sell the landed property is to sell the legal entitlement, which is still latent so long as the usufruct is exercised, but which must one day become active. Here then we really have two owners: one who has enjoyment at present, and the other who will have enjoyment later, but with this difference— that the rights of the first must be exercised in such a way as to reserve the rights of the second for a later date. This is why he may not alter the nature of the landed property which is the very condition on which the right depends. It is said nowadays that the right of usufruct is the outcome of a breaking up of the right of property; it might be more accurate to say that it is the result of a splitting up of this right in the course of time.

There is one thing we must not lose sight of—that the enjoyment in itself is not the distinguishing feature of property. There must be exclusive possession. All other individuals must be prohibited from enjoying the use of the object in question. The right of property consists in essence in the right to withdraw a thing from common usage. The owner may use it or not use it; that is a minor consideration. But its basis in law is to prevent others from using it and even from approaching it. The right of property can be far better defined negatively than in terms of positive content, by the exclusion it involves rather than the prerogatives it confers.

One reservation must be made, however. There is one particular entity which, in certain conditions anyway, can make use of things appropriated by individuals: this is the collective entity represented by the State. The State can in fact, by way of requisition, compel the individual to place the thing owned by him at its disposal. It can even compel him to relinquish it entirely by means of expropriation on grounds of public utility, and the lesser organs of the State, the *communes*, enjoy the same right. It is thus only *vis-à-vis* individuals or private groups that this right of exclusion may be exercised which we have called the distinguishing feature of the right of property. We shall therefore sum up by saying: the right of property is the right of a given individual to exclude other individual and collective entities from the usage of a given thing. The sole exception is the State and lesser organs of the State, whose right of usage can, however, not be exercised except in special circumstances, provided for by law.

This definition will put us on the way to find how the right of ownership came to be established.

Indeed, it follows that the thing appropriated is a thing distinct from common property. Now this feature is also shared by all religious and sacred things. Whenever we have a religious ritual, the world over, the feature that distinguishes the sacred entities is that they are withdrawn from general circulation ; they are separate and set apart. The common people cannot enjoy them. They cannot even touch them. Those who have a kinship, as it were, with sacred things of this kind, can alone have access to them—that is, those who are sacred as they are: the priests, the great, and the magistrates, especially where these latter have a sacred character.

It is these prohibitions that lie at the foundation of what is called taboo, as an institution, that is so common in Polynesia. Taboo is the setting apart of an object as something consecrated, as something belonging to the sphere of the divine. By virtue of this setting apart, it is forbidden to appropriate the object of taboo under pain of sacrilege, or even to touch it. Those alone can have access to it who are taboo themselves or in the same degree as the object. In consequence, there are things which are taboo, forbidden to some but which others may freely make use of. The domain lived in by a priest or a chief was taboo to the common people and could not be made use of by the commonalty, but this very setting apart constituted the full right of property of the owner. Now, although the institution of taboo is especially developed in Polynesia and may be best observed there, it is widely prevalent in other parts. There are only differences of degree between the taboo of the Polynesians and the *sacer* of the Romans. We can see how close the connexion is between this concept and that of ownership. Around the thing appropriated, as around the sacred thing, a vacuum formed. All individuals had to keep at a distance, as it were, except those who had the required qualifications to approach it and make use of it. All about us there are objects it is forbidden to use and, almost, to approach, except for those who fulfil certain conditions: since in the case of taboo, the conditions have religious significance, it is extremely likely that in the other, they will be of the same nature. Therefore we are right in supposing that the origins of property are to be found in the nature of certain religious beliefs. Since the effects are identical, they can in all likelihood be attributed to similar causes.

In some cases, we can observe direct the affinity between the notion of the taboo or the sacred and the notion of ownership. The first gives rise to the other. In Tahiti, kings, princes and the great are sacred beings. Now, the sacred character is in essence contagious: it is communicated to any one who touches the object endued with it. So that a chief can not come in contact with any thing without its becoming taboo by that action, and this in the same degree and the same way as himself. The result is that, *ipso facto*, it becomes his property. In Tahiti, too, personages of this kind go out only when borne on men's shoulders; otherwise they would make the soil taboo and so appropriate it if they touched it with their feet. The connexion between the two ideas is such that very often, one serves for the other. To declare a thing taboo or to take possession of it are one and the same. When a diamond mine was discovered near Honolulu the queen declared it to be taboo in order to reserve the ownership to herself. If a piece of land was made over to a foreigner it was declared taboo in order to protect it from any intrusion by the natives. During the harvest or fishing season, the river or fields were declared taboo to protect the yield. The same applied to the forest during the hunting. "Even ordinary individuals could protect their property by these means. They communicated a sacred character to it or caused this to be done." (Wurtz, VI, 344.) Thus, the taboo ended in becoming a title of possession. Here we see the definite link between taboo and property.[1]

[1] Meaning assumed by Ed.

XIII

THE RIGHT OF PROPERTY (Continued)

W E have seen that the right of property could not be defined by the extent of the rights attaching to the owner. These rights are of two kinds. First, there are the rights of disposal (either by means of alienation or by conversion to another form), which seem more especially characteristic of the right of property. Now, there may be no rights of disposal whatever and yet the right of property will not be extinguished. The minor, the person without civil rights, the person who has a legal guardian, can none of them of their own accord dispose of their property, and yet they remain the owners. On the other hand the family council[1] has the power of disposal within certain limits, without however having any right of property over the thing in question. There remains the power of usage, which, within certain limits, is found wherever there is a right of property. The minor does not have the use of the yield of his property or of the property itself at will, but he has use of it nevertheless in the sense that its yield provides for his upbringing. In this respect there is in fact only a difference of degree between himself and the person who is of age and enjoys his rights in full: the one who is of age, too, cannot make use at will of what he owns, since, if he behaves as a spendthrift he may be deprived of the control of his affairs. All the same, though the power of usage exists wherever there is property, it is not peculiar to it, because it is also to be found in other cases. In particular, anyone may use and use freely, things that are *res nullius*, or those that are *res communes*, that form part of public property, without nevertheless being the owner.

Here we are getting near to what is truly specific in the right of property, if we complete and define this concept of usage by

---

[1] See note, Lect. XII.

adding to it a distinguishing characteristic. One of the features that identify the right of usage which is peculiar to the owner of all similar rights, is that it excludes any concurrent right. Not only does the owner have use of a thing, but he is alone in that use; or, if there are several simultaneous users, it means that there are several owners. Every owner has the right to remove all other individuals from any thing of his own. The way in which he enjoys what belongs to him is of little importance: the main thing is that no other is able to enjoy it in his place. The thing is withdrawn from common usage for his own personal use. It is this, in part, that lies at the root of the idea of appropriation. However, we have not yet grasped the most fundamental element in this notion. Exclusive usage is to be found in many cases where there are, properly speaking, no rights of ownership: I mean those cases where the right of usage is established in a certain way agreed upon as between a given object and one or more given subjects (or individuals), to the exclusion of all others. The right of usufruct is typical of these rights. The proof that this primary feature is already inherent in the right of property is that the usufruct itself is an element in this right; as a rule it is looked on as the result of a breaking up of the right of property. At this stage, then, we have got into the region of things that have to be defined: but we have not yet reached the heart of the matter. There is something that still eludes us. Since the owner can co-exist alongside the usufructuary, this means that the right of usage does not form the whole of the right of property. What does the relation consist in, as between the true or bare owner and the thing owned? It is a moral and juridical bond which makes the status of the thing depend on what befalls the person. If he should die, it is his heirs who inherit. In general, there is a kind of moral community between the thing and the person which makes the one have a share in the social life and social status of the other. It is the person who gives his name to the thing or, conversely, the thing that gives its name to the person. It is the person who raises the status of the thing, or it is the thing, the domain, which—if it has privileges deriving from its origin—transmits them to the person. An entailed estate (or *majorat*) involves the transmission of special rights and a title to the inheritor. If we suppose family inheritance to be abolished as from to-morrow, this bond, characteristic of the right of property, would none the less survive; for there would then be a different kind of hereditary transference: it might, for instance, be the society which would inherit and thus the death of the actual owner would continue to affect the social status of the things owned by him.

These, then, are the two constituent elements of the thing appropriated. We are already aware of the likeness they bear to the sacred thing. The sacred thing is closely akin to the sacred person; it is sacred as this person is and in the same degree. The things that are sacred because they are connected with the head of the religion or State are sacred to a higher degree and by a right different from those connected with sacred dignitaries of lesser rank. The taboo of things is parallel with the taboo of persons. All that modifies the sacred status of the person affects the sacred status of the thing, and vice versa. On the other hand, the sacred thing is set apart and withdrawn from common usage and forbidden to all those not entitled to approach it. It does therefore seem as if the thing appropriated were only a kind of particular species of sacred thing.

There is another resemblance between these two kinds of thing no less characteristic, that shows their fundamental sameness. It is, of course, only another aspect of one of the similarities mentioned. The sacred character, wherever it resides, is in its essence contagious and communicates itself to any object it comes in contact with. Sometimes, when the sacredness is intense, a momentary touch is enough to produce this result: if this is only moderate, there has to be a closer and more prolonged contact. But in principle, all that touches a sacred entity, whether person or thing, becomes sacred in the same way as this person or thing. The potency that is within the sacred entity (and which renders it sacred) is seen in the popular imagination as ever ready to spread into all the *milieux* open to it.

Indeed, it is partly from this fact that the ritual interdictions derive which separate the sacred from the profane; it is a matter of insulating this potency, of saving it from being lost or dissipated or from disappearing. This is why I said that the contagious quality is only another aspect of the insulation that is a feature of sacred things. Again, since the sacred feature, by thus communicating itself, makes the objects to which it is transmitted enter the region of sacred things, we might say that, as a rule, the sacred draws to itself the profane with which it is in contact. How this strange phenomenon arises it is useless to pursue here, all the more so since there is no satisfactory explanation. The truth of the fact is, however, not in doubt; we have only to refer to the instances of contagiousness in the taboo that were described in the last lecture.

The characteristic that makes a thing the property of a certain subject or individual exhibits the same contagiousness. It tends always to pass from the objects in which it resides to all those objects that come in contact with them. Property is contagious. The thing appropriated, like the sacred thing, draws to itself all things that touch it and appropriates them. The existence of this singular capacity is confirmed by a whole collection of juridical principles which the legal experts have often found disconcerting: these are the principles that decide what is called the 'right of accession'. The idea may be expressed in this way: any thing to which another of less importance is added (*accedit*) communicates to it its own status in law. An ownership that comprises the first is extended *ipso facto* to the second and comprises it in turn. This one in turn becomes the thing owned of the same proprietor as the other. Thus, the fruits, the yield of the thing belong to its owner, even though they are separate from it. By virtue of this principle, the young of animals belong to the owner of the mother; the same rule applies to slaves. This is because there is immediat contact between mother and young and not between them and the father. In the same way, all that the slave earns belongs to the estate to which he is bound, to the master who owns this estate. The son of the family was, as we have heard, owned by the head of the family. The rights of the head of the family are extended by contagion from the son to all that the son earns. I may build a house with my own materials on the estate of another, and the house becomes the property of the owner of the estate. He may be bound to give me compensation, but the right of ownership accrues to him. It is he who enjoys the tenure of the house, and if he should die, his heirs would inherit. The deposits of alluvium along my estate are added to this property and my right of ownership is extended to the things thus added. The proof that this is a matter of contagion by contact is, that where there is separation, when a boundary is made to a field and it is thus detached in law and psychologically from its surroundings, the right of accession does not arise. In the same way, when the trees of my neighbour send out roots beneath land that I own, common possession is established and my rights of ownership are extended to these trees.

In all cases, it is the more important thing that draws the lesser to it. This means that, as the two rights of ownership are in conflict, it is of course the one with the greater force that exerts the greater power of attraction. Not only does the right spread in a general way, but it spreads, at the same time keeping the very same specific characteristics. For instance, the estate that passes by inheritance is, in many societies, inalienable. This inalienability is carried over from the estate to objects most constantly connected with the estate, that is, beasts of burden or draught animals. And the proof that this second inalienability derives from the first is, that it can be extinguished at an earlier stage and more easily. Indeed, there are many rights in which the traces of the inalienability of the land or buildings still exist, whilst all recollection of the inalienability of the agricultural implements has been lost.

Thus, we find on every hand striking analogies between the idea of the sacred thing and that of the thing appropriated. Their characteristics are identical. Further, we have seen that, in fact, the communication of the sacred character very often brings about an appropriation. To consecrate is a way of appropriating. What indeed does 'to consecrate' mean, if not to appropriate a thing to a god or consecrated personage and to make this thing his own? Let us imagine some kind of symbol of honour and merit subject to the use of ordinary people and available to everybody : it is already plain that there is little difference between this and a form of appropriation. But although what we have just discussed leads us to admit that such consecration may be possible, it still remains to show its reality.

To do this, we must examine the most ancient form of property known, that is to say, landed property. It is only from the time when agriculture became established that this kind of property can really be closely examined. There existed until then only a vague right of all members of the clan over the whole area held by them. A clearly defined right of property only appears at the focal point of the clan; small family groups settle on agreed pieces of land, set their landmark on them and dwell there permanently. Now it is certain that this ancient family holding was permeated by a profound sacredness and that the rights and privileges associated with it were of a sacred kind. Indeed, a proof of this is the fact that it was inalienable. For the inalienability has the distinctive characteristic of the *res sacrae* and of the *res religiosae*. And what indeed is inalienability, if not an insulation or setting apart more complete and more radical than that involved in the exclusive right of usage? An inalienable thing is one which must belong always to the same family, that is, not only at the present but in perpetuity, and which is withdrawn from common use. Not only is it impossible for individuals outside the family to enjoy this thing in the present, but they can never do so. The boundary separating them from the thing can never be crossed. We can see that, in some ways, the right to alienate or to sell is far from being the highest point of development in the right of property: it is rather the inalienability itself that represents this point. For it is here that appropriation is most complete and best defined. It is here that the bond between the thing and the subject (or individual) who is the possessor reaches its maximum force and here too that the exclusion of the rest of the society is most strictly imposed.

But this sacred nature of landed property is revealed in its very structure. The customs we shall go on to discuss have been especially observed with the Romans and the Greeks and in India. But in any case they were certainly very widespread.

Each field or holding was surrounded by a belt separating it clearly from all the neighbouring holdings, private or public. This was a strip of land a few feet wide that had to remain uncultivated and untouched by the plough. (Fustel de Coulanges.) Now, this belt was sacred—it was a *res sancta*. This was the name given to things which, without being, strictly speaking, *divini juris*, in the sphere of the gods, were so, however, to a certain extent, *quodam modo*, as Justinian says. To violate this sacred surrounding belt, to till it and profane it amounted to sacrilege. The man who committed such a crime was accursed, that is, declared *sacer*, both himself and his oxen, and therefore anyone might kill him with impunity. "He was condemned to have no issue and his race to death; for the extinction of a family, in the eyes of the ancients, was the supreme vengeance of the gods."

We know, moreover, the ritual ceremony by which the sacred character of this space was maintained in a regular way. "On certain appointed days of the month and year, the head of the family made a tour of his field following this line; he drove the victims before him, singing hymns and offering sacrifices." (Fustel de Coulanges.) It was the path followed by the victims and sprinkled with their blood which constituted the inviolable boundary of the holding. The sacrifices took place on great stones or trunks of trees set up at intervals and called terminals. Let us hear how Siculus Flaccus described the ceremony. "This", he says, "was the practice of our ancestors: they began by digging out a little ditch and setting up the terminal stone on the edge, they crowned it with garlands of herbs and flowers. Then they offered a sacrifice; the victim slain, its blood was made to flow into the ditch, and into this they threw live coals, grain, cakes, fruits, a little wine and honey. When all these had been consumed in the ditch on embers still glowing, the stone or block of wood was driven in." It was this sacred act that was repeated every year. The terminal or boundary stone thus assumed a character that was decidedly sacred. With time, this sacred character became personified, and hypostasized in the form of a definite deity: this was the god Terminus, whose various terminals, placed about the field, were considered, in a sense, as so many altars. Thus, once the terminal was set up, no power on earth could displace it. "It had to remain in the same place for all eternity. In Rome, this sacred principle was expressed in a myth: Jupiter, desiring to have a site for a temple on the Capitoline Hill, had not been able to dispossess the terminal god. This ancient tradition shows to what degree property was sacred, for the immovable terminal means no more than inviolable property." These ideas and practices were moreover not peculiar to the Romans. To the Greeks, too, the boundaries were divine and became Θέοι ὅροι. We find the same ceremonies for making boundaries in India. (*Manou,* VIII, 245).

It was the same with gates and walls. "*Muros sanctos dicimus quia poena capitis constituta sit in eos qui aliquod in muros deliquerunt.*" It is believed that the phrase related only to the gates and walls of cities. But this limitation is arbitrary. The boundary wall of all houses is sacred: ἕρκος ἱερὸν, said the Greeks. In many countries, it was on the threshold that this sanctity reached its greatest force. Hence the custom of lifting the betrothed over the threshold before bringing her in, or of making an expiatory sacrifice on it. This, because the betrothed is not of the house. She is committing a kind of sacrilege in treading on sacred soil, which, if not prevented, has to be expiated. Further, it is usual for the building of a house to be accompanied by a sacrifice similar to the one for the marking of the field boundary. The purpose of this sacrifice was to sanctify either the walls or threshold or both at once. Sacrificial victims were immured within the walls or in the foundations; or they were buried beneath the threshold. Hence, its sacred character. It was a ritual similar to the one for marking the boundaries of a city. These solemn ceremonies were well known: the myth of Romulus and Remus carries on the memory. They served for private houses as well as for public domains.

So there is a sacred basis for property being property. It consists, in fact (following on what we have said), in a kind of insulation of the thing, which withdraws it from the common area. This insulation has sacred origins. It is the ritual procedure that creates—on the confines of the field or around the house—an enclosure that in each case makes them sacred, that is, inviolable, except for those who conduct these ceremonies, which means the owners and all that belongs to them in the way of slaves and animals. What amounts to a magic circle is drawn about the field, which shields it from trespass or encroachment, because such intrusions, in these circumstances, become sacrilege. But although it is clear that these practices led to the appropriation of the thing thus insulated, we may still not see how the practices themselves originated. What are the ideas that brought men to carry out these rites and so to yield to the gods the fringes of their domains and to make them sacred ground? It is true there might be a very simple answer : that these customs were only expedients resorted to by individuals to enforce respect for their possessions. The owners might have made use of religious beliefs to keep intruders at a distance. But a religion does not descend to the level of expediency, unless the beliefs it inspires are no longer a very living thing. The customs we have described are far too primitive to have been expedients intended to safeguard worldly interests. Moreover, they were as much a source of constraint as an advantage to the owners, for they fettered their freedom of action. For these customs did not allow them to alter the boundaries of the holding or to sell it if they wished to. Once consecrated, even the master himself could do nothing to change the enclosure in any way. It was, then, an obligation he was under, rather than an expedient invented by himself in his own interest. If he adopted the procedure we describe, it is not because it was useful to him but because he had to act in this way. (Terrible nature of some of these sacrifices, that of a child). But what are the causes that lie behind this obligation?

Fustel de Coulanges thought they lay in the cult of the dead. Every family, he said, has its own dead; these dead are buried in the field. They are sacred beings, for death makes almost gods of them—and this nature is therefore extended to the ground in which they lie. This ground becomes theirs by the very fact that they lie in it, and it is thereby sacred. We can understand that this sacredness spread from the little mound serving as a family sepulchre to the whole field. The inalienability of property established in this way is thus explained. For the true owners of this domain are divine beings and their right is indefeasible. The living are not free to dispose of the field because the right does not lie with them.

There is no doubt that these burial places were especially sacred. They could not be sold. And although Roman law allowed a family to sell its field or holding (such a sale was difficult and met with all kinds of obstacles) it had always to remain owner of the tombs. Does this mean that the right of property is only an extension of this sanctity of the tomb? The theory is open to a number of objections:

(1) If, at a pinch, it may explain the holding of the field as

property, it does not account for the holding of the house as property. For the dead were not given burial in both of these places. It is true that Coulanges does not avoid this paradox. When he explains the sanctity of the hearth, he imagines that formerly the ancestors were buried beneath the hearth-stone, and when he explains why the field is sacred, he infers the presence of the dead at the heart of the field. They could however not be in both these places.

(2) The facts on which his assumption rests that the dead were buried in the field, are few and unconvincing. There is no proof surviving in Latin and the few texts cited are very inconclusive. In any case, this custom was not nearly so widely accepted as the sacredness of landed property, inviolable and inalienable.

(3) What is more decisive, the very siting of the sacred area in the field is against this explanation. If the focal point of this sanctity were the burial place, it is there that it would attain its greatest force and, inevitably, it would diminish towards the confines. But it was on the periphery, on the contrary, that it was most intense. This is where the belt or strip of land was, reserved to the terminal god. So that it was not the family tomb the strip protected, but the whole field. If the only aim was to insulate the ancestral tombs, it is around these tombs and not at the outer fringes of the domain that this line of insulation would have been drawn.

This error of Fustel de Coulanges springs from too narrow a concept of the family cult. He reduces it to the cult of the dead, when in reality it is far more complex. Family religion was not ancestor worship alone; it was the cult of all things that played a part in the life of the family, such as the harvesting, the seasonal fruitfulness of the fields, and so on. If we take this overall view, the practices we have described become intelligible. We have to remember that, from a certain point in evolution, the whole of nature takes on a sacred character—πάντα πλήρη Θεῶν, gods crowd in everywhere. The life of the universe and of all things in it is related to an endless stream of divine principles. The fields, until then uncultivated, become inhabited, possessed by sacred beings conceived in some form, personified or not, who have command over them. Like all else in the world, the field has a sacred character. This quality makes it unapproachable. It matters little whether these sacred beings are devils malign by nature, or deities on the whole benevolent. The husbandman cannot enter his field without trespassing on their domain; he cannot till or shift the soil without disturbing them in their possession. Thus, he is exposing himself to their anger, which is always redoubtable, if he does not take the right precautions.

All this being granted, the rites we have described seem remarkably similar to others, well-known, that throw light on them: these are the sacrifices of the first-fruits. Just as the soil is a divine thing, the harvest which is the fruit of this soil contains, too, a principle that is divine. In the seed sown in the earth there is a sacred force that develops in the shoots of the corn and reaches its fulfilment in the grain. Thus the grains of corn are also sacred, since they have a god within them and are this god made manifest. In consequence, mortals may not touch them until certain ritual ceremonies have tempered the sacredness that resides in them, and in such a way that they can be made use of without peril. It is this purpose that is served by the sacrifice of the first-fruits. The supreme and most formidable element in this sacredness is concentrated in a sheaf or a number of sheaves, usually the first sheaves garnered, and these are sacred: no one touches them and they belong to the god of the harvest; they are offered to him and no mortal dare partake of them. The remainder of the harvest, although still keeping a certain sacred quality, is rid of something that made an approach to it too perilous. This remainder can serve for everyday use without the user being exposed to divine vengeance, for the god has had his due, and this came to him solely because the element in the harvest that was too sacred had been got rid of. The sacred element residing in the crops has been prevented from passing over into the profane, for it has been separated from the profane, and by the sacrifice, it has been kept within the divine sphere. The line of demarcation of the two worlds has been respected, and this is the supreme sacred obligation. What we have said on the harvest applies equally to all fruits of the earth. Hence the rule prohibiting men from touching these fruits, whatever they be, without having set aside the first-fruits beforehand and offered them to the gods. There is no religion that is not familiar with this practice.

The analogies with the ritual ceremonies of the boundary stones are striking. The field is sacred, it belongs to the gods, therefore it may not be used. To enable it to serve profane ends, recourse is had to the same procedure as used in the harvest or the vintage. It has to be relieved of the excess of sacredness in order to make it profane or at least profanable, without incurring peril. The sacredness, however, is indestructible: it can therefore only be shifted from one point to another. This dreaded force dispersed about the field will be drawn off, but it has to be transferred elsewhere, so it is accumulated at the periphery. This is the purpose of the sacrifices described. It is upon an animal that the diffused sacred forces are now concentrated: then this creature is led all around the field. Wherever it passes, it communicates to the soil it treads on the sacred character that is in it and which it has drawn off from the field. This ground becomes sacred. In order to fix this awe-inspiring sacredness upon it the better, the animal is sacrificed and in the very ground furrowed for the purpose the blood of the victim is made to flow, for the vital fluid is the supreme vehicle of all sacred principles. The blood is life itself, the living thing. Henceforward, the strip that was the scene of this ceremony is consecrated land: what was divine in the field is now transferred to the strip. Moreover, it is set aside and may not be touched; it is not to be tilled and cannot be changed in any way. It does not belong to men but to the god of the field. From now onwards all within the domain is at the disposal of men to make use of for their own needs. By the very fact, however, that the sacredness has been, as it were, pushed back to the boundary of the land, this land is *ipso facto* fenced in, surrounded by a circle of sanctity protecting it against any intrusions or occupancy from without.

Furthermore, the sacrifices made in these circumstances are quite likely to have had more than one purpose. Since the husbandman, in spite of all, had interfered with the possession of the gods, and committed an offence that placed him in peril, it was necessary to atone for it. The sacrifice brought about this atonement at one stroke. The victim took upon itself the offence committed and expiated it on behalf of the guilty. And then (as a further result), owing to the ritual carried out, not only were the deities disarmed but transformed into protective

powers. They kept guard over the field, defended it and ensured that it should prosper. We could interpret the practices
that were current when a house was built in the same way. To
get this house built has meant disturbing the guardian spirits
of the soil. They have been roused and antagonized. Thus,
any and every house is forbidden to us and taboo. Before
entering, there must first be a sacrifice. The victims are killed
on the threshold or on the foundation stones. By this means the
guilt of sacrilege was atoned for. At the same time the vengeance which has threatened is changed to favour and the angry
dæmons are turned into protective spirits.

But only those who had carried out the ritual could make use
of the field or house. They alone have atoned for the sacrilege
committed; they alone have conciliated the divine elements
with whom they are in communication. The deities had an
absolute right over the things: those who conciliated them have
in part taken their place in so far as this right over things is
concerned. It is those alone who effected this process of sub-
stitution who can benefit by it. They alone, therefore, can exercise the right won, as it were, from the gods. The power to use
and make use of belongs to them exclusively, in their own
right. Before the ritual was carried out, everyone had to keep at
a distance from the things entirely withdrawn from profane
use; afterwards, everyone was bound to the same stricture,
these others alone excepted. The sacred virtue that until then
protected the divine domain from any occupancy or trespass,
was henceforth exercised for their benefit: it is that virtue
which constitutes the right of property. It is because they have
enlisted its service in this way that the land has become theirs.
A moral bond has been forged between themselves and the
gods of the field by the sacrifice, and since the link already
existed between the gods and the field, the land has therefore
become attached to men by a sacred bond.

This, then, seems to show how the right of property had its
origin. Man's right of property is only a substitute for the right
of property of the gods. It is because the things are by nature
sacred, that is, appropriated by the gods, that it has been
possible for them to be appropriated by the profane. Also,
the quality that makes property an object of respect and
inviolable—a quality which in effect makes it property—is not

communicated by men to the domain; it is not an attribute which has been inherent in men and from them has devolved on things. It is in things that the quality originally resided, and it is from things that it has risen towards men. The things were inviolate in themselves by virtue of sacred concepts, and it is this derived inviolability that has passed into the hands of men, after a long process of being diminished, tempered and canalised. Respect for property is, then, not—as we often hear —an extension to things of the respect that human personality imposes, individually or collectively. It has a very different source, exterior to the human person. If we want to know whence it comes, we have to see how things and men acquired a sacred character.

XIV

THE RIGHT OF PROPERTY (Continued)

PROPERTY is property only if it is respected, that is to say, held sacred. We might think *a priori* that this sacred character derived from man—that it is the husbandman who has communicated to the soil he tills and works, something of the respect of which he himself is the object, of the sanctity which is in him. In this case the property would have no moral value except that lent to it by human personality: this would be the value which, by entering into a relation with things and by making them its own, would confer a certain dignity on them by extension, as it were. But the facts seem to prove that the notion of property came about in quite a different way. The kind of sacredness that kept at a distance from the thing appropriated all individuals except the owner, does not derive from the owner; it resided initially in the thing itself. The things were sacred in themselves; they were inhabited by potencies, rather obscurely represented, and these were supposed to be their true owners, making the things untouchable to the profane. The profane were therefore not able to intrude on the divine sphere, unless they gave the gods their due and expiated their sacrilege by sacrifices. With these preliminary safeguards, they were able to take over the right of the gods themselves and put themselves in their place. Although, thanks to this expedient, the sacred character of the field ceased to be a hindrance to the work of the husbandman, it had not become extinct. It had merely been shifted from the centre to the periphery, and there its natural potency worked against all those who had not acquired a kind of immunity to it. The gods had not been driven from the field but transferred to the confines: a kind of bond had been made between them and the owner; they had become his protectors and by these regular ceremonies he ensured that their favour should continue.

But for all those outside, they were still powers to be dreaded. Woe to the neighbour whose plough had so much as grazed a terminal god! They had disarmed only towards those who had paid the debt due and had behaved to them in a proper manner. The field was in this way shielded from any incursion or from any seizure by another. A right of property became established for the benefit of particular men. This right has, then, a sacred origin: human property is but sacred or divine property put into the hands of men by means of a number of ritual ceremonies.

We might perhaps be astonished to find an institution so fundamental and widespread as property thus resting on illusory beliefs and ancient notions which are held to have no objective foundation. Guardian spirits of the soil or the fields do not exist, we may say; how then has a social institution been able to persist, if it rests on fallacies alone? It should have crumbled away, it might seem, as soon as it came to be realized that these mystic concepts were utterly empty. But it happens that religions, even the most uncouth, are not, as is sometimes believed, merely phantasies that have no basis in reality. Certainly they do not express the things of the physical world as they are; they have little value in throwing light on the world. But they do interpret in a symbolic form, social needs and collective interests. They represent the various connexions maintained by society with the individuals who go to make it up, as well as the things forming part of its substance. And these connexions and interests are real. It is through a religion that we are able to trace the structure of a society, the stage of unity it has reached and the degree of cohesion of its parts, besides the expanse of the area it inhabits, the nature of the cosmic forces that play a vital role in it, and so on. . . . Religions are the primitive way in which societies become conscious of themselves and their history. They are in the social order what sensation is in the individual. We might ask why it is these religions distort all things as they do in their processes of imagery. But is it not true that sensation, equally, distorts the things it conveys to the individual? Sound, colour and temperature have no more positive existence in our world than the gods, the dæmons or spirits. By the fact alone that the representation presupposes a subject represented—(here individually and there collectively)—the nature of this subject is a factor in the representation and alters the shape of the thing represented. The individual, in picturing by means of sensation the relations he has with the world about him, puts into these images something that is not there, some qualities that come from his own mind. The society does the same thing in picturing by means of religion the *milieu* that constitutes it. The distortion, however, is not the same in both instances, because the subjects differ. It is for the thinkers to rectify these illusions that are necessary in practice. We may at any rate rest assured that the religious beliefs we find at the base of the right of property conceal social realities which they express in metaphor.

To make our interpretation really convincing, we have to get through to the realities and to discover beneath the letter of the myths the spirit it expresses. That is, we have to perceive the social causes that gave rise to these beliefs. The question comes back to this: how is it that the collective imagination has been led to consider the soil as sacred and inhabited by divine principles? The problem is far too wide in scope to be treated here, all the more so since the solution still escapes us. There is, however, a way of forming some image of things that will serve our purposes—one that will allow us to see how the illusions that come from a region of myths can have in reality a positive significance.

The gods are no other than collective forces personified and hypostasized in material form. Ultimately, it is the society that is worshipped by the believers; the superiority of the gods over men is that of the group over its members. The early gods were the substantive objects which served as symbols to the collectivity and for this reason became the representations of it: as a result of this representation they shared in the sentiments of respect inspired by the society in the individuals composing it. This is how deification came about. But although the society is superior to its members taken singly, it exists only in them and through them. The collective imagination therefore had to be brought to the point of conceiving of sacred beings as indwelling in men themselves. This is indeed what happened. Every member of the clan is supposed to carry within himself a share of the totem whose cult is the religion of the clan. In the Wolf

Clan, each individual is a wolf. There is a god within him and indeed several. If, then, there are gods in things and especially in the soil, it is because things, and especially the soil, are associated with the intimate life of the group just as much as human beings are. This is because they are believed to live the life of the community. Therefore it is quite natural that the principle of communal life should reside in them and make them sacred. We now get an idea of what this sacred character is, that the soil is imbued with. It is not a mere invention without foundation, some figment of a dream. It is a stamp the society has put on things, because they are closely mingled with its life and form part of itself. If the soil was not to be approached by the foot of individuals it is because it belonged to the society. This is the true potency that set it apart and withdrew it from any private appropriation. To sum up, we might say: that private appropriation pre-supposes an initial collective appropriation. We have said that the believers took upon themselves the right of the gods; we should now say that the individuals took upon themselves the right of the collectivity. It is from this collectivity that all sacredness issued. It alone (if we confine ourselves to things empirically known) has adequate power to raise the existent thing—whether it be land, animal or person— above and beyond the reach of any private assault. Private property came into existence because the individual turned to his own benefit and use the respect inspired by the society, that is, the higher dignity with which it is clothed and which it had communicated to the things composing its material substitute. As to the hypothesis, according to which the group was the original possessor of things, that fits in perfectly with the facts. Indeed, we know that it is the clan that owned the land in common, land that it was settled on and which served for hunting or fishing.

Looked at from this point of view, even the ritual practices we have described take on a new significance and can be defined in secular terms. The sacrilege that man thinks he is committing against the gods by the very fact of tilling and breaking up the soil, is in truth committed against society, since society is the reality hidden behind these mythical concepts. It is therefore, in a way, to society that man makes his sacrifices and offers up the victim. Again, when these figments of men's minds dissolve, when these phantom deities vanish into air and the reality they represent appears by itself alone, it is to this society that these annual tributes will be offered, by which the believer originally bought the right from his deities to till and cultivate the land. These sacrifices, these first-fruits of all kinds, are the earliest form of taxes. First, they are debts that are paid to the gods; they then become tithes paid to the priests, and this tithe is already a regular tax that later on is to pass into the hands of the lay authorities. These rites of atonement and propitiation finally become what amounts to a tax, although unsuspected. The germ of the institution is there, however, and is destined to develop in the future.

If this interpretation is right, the sacred nature of appropriation had for a long time simply meant that private property was a concession by the collectivity. But however this may be, the circumstances in which property came into being did determine its nature. It could only be collective. In fact, it was by groups that the land was appropriated in this way, that the formal ceremonies described were carried out and thenceforth the whole group had the benefit of the results. These formalities even had the effect of giving the land a personal identity and cohesion that it did not originally possess. This strip of consecrated land separating the field or holding from those adjoining, at the same time insulates all those within it from similar groups settled elsewhere. This is why the coming of agriculture undoubtedly gave to family groups smaller than the clan a cohesion and stability they had not known before. It was truly the individual nature of the field that made the collective individuality of these family groups. Henceforward, these groups no longer yielded to the slightest change in their circumstances: no longer did they take shape for a time and then disperse, according to the impulse of private sympathies or fugitive interests. They possessed a definite form, a bone structure, as it were, which made its indelible pattern on the very land they lived on: for it was indeed they who made the form and contour itself of that land—a form that was unchanging.

This goes to explain one of the features of collective family property already referred to in last year's lectures. It means that under this system people are possessed by things at least as much as things are possessed by people. Kindred are kindred only because they make common use of a certain domain. If anyone makes a final parting with this economic community, all links of kinship with those that remain are cut. This predominating influence of things becomes very clear from the fact that in some circumstances people may leave the group thus formed and cease to be kindred. The things, on the contrary, the landed property and all that goes with it, remain there in perpetuity, since the patrimony is inalienable. In some cases, this possession of people by things goes so far that it ends up by becoming a real form of slavery. This is what happened to the 'epicleros' daughter in Athens. If the father had as offspring a daughter only, she would inherit, but it was the status in law of the property which came to her that fixed her own status in law. Since the estate could not go out of the family, precisely because it was the very heart of it, the heiress was bound to marry her nearest male relative; if she was already married, she had either to break her marriage or renounce her heritage. The person followed the thing: it was a question of the daughter being inherited rather than inheriting. All these facts are easily explained if landed property has indeed the origin we have assigned to it. For then it is the property in the form of real estate that binds the land to the family; it is that property that has made the family's centre of gravity and that has even imparted to the family its own external forms. The family means the individuals taken as a whole who lived in this insulated and sacred little island that made up the domain. It is the laws that bind them to the sacred soil they cultivated which therefore unite them amongst themselves. This then, generally speaking, is how the kind of cult whose object is the family field or holding came into being, and the sacred prestige and awe this cult inspired in men's minds. The cult did not acquire this prestige merely through the vital importance of the soil to the husbandman, nor from the supreme power of tradition, but simply because the soil itself was steeped in sacred meaning. It was far more a case of the sacredness of the holy thing being communicated to the family, than of its deriving from the family.

But precisely because property, in its origins, can only be collective, it remains to explain how it became something individual. How is it that individuals thus grouped together, attached to an identical group of things, came to acquire separate rights over separate things? The land holding cannot, in principle, be broken up: it forms a single unit, and that is the unit of the inheritance; and this indivisible unit is imposed upon the group of individuals. How does it happen that, in spite of this, any individual should have been able to reach the point of having a property of his own? As we might guess, this individuation of property could not come about without involving other changes in the situation as between things and persons. For as long as the things preserved this moral superiority over persons, as it were, it was impossible for the individual to become their owner and establish his own command over them.

There were two different causes underlying this result. To begin with, it was enough for one of the members of the family group to be raised in rank in some way—by a chain of circumstances—for him to be lent a prestige that none of the others had and to make him the representative of the family group. In consequence, the ties binding the things to the group bound them direct to this privileged personality. And since this individual embodied in himself the whole group, men and things, he was in fact invested with an authority that placed things as well as men under his dominance, and thus an individual property came into existence. This change was achieved with the coming of paternal, and more especially, patriarchal, power. We heard last year what the causes were that led the family to emerge from a state of close unity and equal rights—still seen until recently in Slav families—and to elect a head to which it submitted itself. We saw how, by that very fact, this head of family became a high moral and sacred power: this is because the whole life of the group was absorbed in this head, and thus he came to have the same transcendence over each of its members as the collectivity itself. He was the family entity personified. And it is not alone people, traditions and sentiments that happen to find expression in his person. It is, too, above all the patrimony, with all the concepts attaching to it. The Roman family was made up of two kinds of elements: the head of the family, on the one hand, and on the other the rest of the family, called the *familia*, which comprised at once the sons of the family and offspring, the slaves and all things or property. All that was of moral or religious significance in the *familia* was, as it were, concentrated in the person of the head of the family. This is what gave him such a supreme position. The family's centre of gravity thus became displaced. It passed from the things it was vested in to a given person. Henceforward an individual came to be an owner, in the full sense of the word, since the things were subject to him, rather than he to them. It is true that so long as the authority of the head of the family was as absolute as it was in Rome, he alone could exercise this right of property. But when he had passed away, each of his sons, successively, was called upon to exercise it in turn. And by degrees, as the patriarchal power became less despotic, at least as a right, and as the individuality of the sons came to be acknowledged even before the death of the father, they were able—to some extent, at any rate—to become owners in his lifetime.

The second cause of the individual becoming an owner was no less effective in the result. Its action ran parallel to the effects of the first that I have just described and it reinforced them.

This second cause was the development in the sphere of personal or movable property. Indeed it was only landed property that had the sacred character. This had the effect of withdrawing it from being within the disposal of individuals and so made a communal system necessary. Personal or moveable property, on the other hand, was in itself, as a rule of a profane nature. However, so long as industry remained solely agricultural, personal property played only a secondary and auxiliary part; moveables were hardly more than adjuncts or annexes of landed property. This was the centre to which all that was moveable in the family gravitated, things as well as people. It kept all *things* within its sphere of action and thereby prevented them from acquiring any legal status in keeping with their particular features, and from developing the germ within them of some new right. Also, any earnings that members of the family could make outside the family community flowed into this family patrimony and was merged with the rest of the property, on the theory that the accessory follows the principal. But as we said, the implements and the dead or live stock that were used more especially in farming and were therefore in closer contact with the soil, shared with it its characteristic attribute; that is, they were inalienable. With time, however, and with the progress of trade and industry, the personal or moveable property took on greater importance; it then cut away from this landed property of which it was only an adjunct; it played a social rôle of its own different from landed property, and became an autonomous factor in economic life. Thus a fresh nucleus of property was made outside real estate, and so did not of course have its characteristic features. The things comprised in such a nucleus had in themselves nothing that put them beyond the reach of any trespass such as we discussed. They were only things, and the individual into whose hands they came was likely to find himself on an equal footing or even above them. He could therefore dispose of them more freely. Nothing tied them to any given point in space; nothing made them immoveable. This meant that they depended direct only on the person of the one who acquired, or on some way in which he had acquired them. And that is how this new right of property came about. But it is clear, in the light of our present-day laws, that real estate and moveable property are quite different in nature, and this reflects the separate phases of evolution in the law. The former is still loaded with prohibitions and obstacles which are mementos of its ancient sacred character. The latter has always been freer, more flexible, more entirely left to the discretion of individuals. Real as this duality may be, we must not lose sight of the fact that the one type of property issued from the other. Personal property, as a distinct entity in law, was formed only as a result of landed property and on its pattern: it is a weak reflection, an attenuated form of it.

It was landed property as an institution which first established a bond *sui generis* between groups of persons and certain given things. Once that had been done, public opinion was quite naturally ready to admit that, as social conditions changed, bonds similar in the main might link things with personalities in place of collectivities. This was only applying a previous system of regulations to new circumstances. Personal or moveable property is in a way no other than immoveable property modified to accommodate the features peculiar to

167

moveable property. It bears the stamp of its origins even to-day. It may in fact be inherited and by the same title as the other; in the case of descent in direct line, the right of succession has to be observed. Inheritance is undoubtedly a survival of the phase of early communal property. It seems to be a fact that this communal property, which in the beginning was identified with real property or immoveables, was in reality the prototype of personal or moveable property.

It is now clear how property as we know it to-day is linked with the mystic beliefs we have found at the root of the institution. Originally, property was related to land, or at least the distinguishing features of landed property extended even to moveables, owing to their lesser importance; these features, by virtue of their sacred nature, imply of necessity communalism. Here, then, we have the starting point. Then, by a dual process, individual ownership splits off from collective ownership. The concentration of the family, on the other hand, which established patrimonial powers, causes all these sacred virtues (that were inherent in the patrimony and gave it an exceptional status), to issue from the person of the head of the family. From now onwards, it is man who stands above *things*, and it is a certain individual in particular who occupies this position, that is, who owns or possesses. Whole categories of profane things take shape independently of the family estate, free themselves of it and thus become the subject of the new right of property, one that is in its essence individual. Then again, the individualising of property followed from landed property losing its sacrosanct quality—a quality which was absorbed by the human being. It was due also to the fact that the other form of property, which in itself did not have this quality, evolved to the point of having a distinct and different juridical structure. But since communal property is the stock from which the other forms sprang, we find traces of it in their structure as a whole.

It may appear surprising to see no part assigned in the origins of the right of property to the concept of its deriving from labour. But if we look at the way in which the right of property is regulated by our code of law, we shall not find this principle expressly laid down in any part of it. Articles 711 and 712 of the Code of Civil Law say that property is acquired by inheritance

or succession, by gift or donation, through accession, by uninterrupted possession, or by the effect of binding obligations.
Of these five methods of acquisition, the first four do not imply
the concept of labour in any way,[1] and the fifth not necessarily
so. If a sale transfers the ownership of a thing to myself, it is
not because this thing has been produced by the labour of the
individual making it over to me, nor because what I give in
exchange is the result of my labour; it is simply because both
the one and the other are in the lawful possession of those who
exchange them and this possession is founded on a valid right.
In Roman law, there is even less evidence of the principle.
We might say that in law, the vital element in all methods of
acquiring property is : the material taking possession, the
holding of it and the close contact with it. Not that this physical
fact is enough to constitute ownership; but it is always necessary, at least initially. Furthermore, what demonstrates *a priori*
that this concept has not affected or, at least, deeply affected
the right of property is that it is of quite recent origin. It is not
until we get to Locke that we see the theory that property is
legitimate only if it is founded on labour. Grotius seemed still
unaware of it at the beginning of the century.

Is this to say that the theory does not appear in our laws?
Not at all; but it did not derive from any provisions relating
to the right of property. It is to the right of contract that we
have to look. Moreover, it seems right to us that all work that
is or can be put to use by others should have remuneration and
that this should be in ratio to the useful labour expended. All
remuneration confers rights of ownership, since it transfers
things to the beneficiary. By this means a change or transfor-
mation in the right of contract has come about, which must of
necessity affect the right of property. It may even occur to us
that the principle which was in process of evolving is in conflict with the principle on which any personal appropriation
has hitherto rested. For we cannot have just work by itself:
it calls for some material substance, some object it has to be
applied to, and this object must have already been appropriated, since the work is done to modify it. The work therefore does away with the appropriations that are not founded on

---

[1] Tr. note : accession may include the application of labour, acc. to A. W.
Dalrymple (Legal Terms.)

work. Hence these conflicts between the new demands of conscience which are beginning to set in, and the earlier concept of the structure of the right of property. But since these new demands have their origin in the new concepts that we begin to see in contractual law, it is proper to examine them in the principles of contract.

XV

THE RIGHT OF CONTRACT

WE have seen the way in which the right of property appears to have been established. The sacredness diffused in things, which withheld them from any profane appropriation, was conducted by means of a certain definite ritual either to the threshold or to the periphery of the field. It there established something like a girdle of sanctity or sacred encircling mound, protecting the domain from any trespass by outsiders. To cross this zone and enter the little island insulated from the rest of the land by ritual, was reserved to those alone who had carried out the rites, that is, those who had contracted especial bonds with the sacred beings, the original owners of the soil. By degrees, this sacredness residing in the things themselves passed into the persons: the things ceased to be sacred in themselves; they no longer possessed this quality, except indirectly, because they were subject to persons who themselves were sacred. Property, from being collective, became individual. For, so long as it derived exclusively from the sacred quality of objects, it was not related to any definite subject or individual; it was not in persons, nor (with all the more reason) in any particular person that it had its source and place of origin, and therefore no person whatever could be considered as holding it. The whole group enclosed by this, as it were, sacred zone had equal rights, and the fresh generations were destined to enjoy the same rights by the very fact of being born within that group. Individual property came into being only when an individual split off from the family aggregate who embodied in himself all the sacred life diffused amongst the people and things of the family, and who became the holder of all the rights of the group.

It may be surprising to see the right of individual property thus linked to sacred concepts of ancient times, and we might be inclined to think that representations of this kind could not amount to a very solid foundation for this institution. We have, however, already seen that although religious beliefs are not based on fact, they do nevertheless express the social realities, even when they interpret them by symbol and metaphor. We know indeed that the sacred character which still marks the individual to-day is founded in reality; it is no more than the expression of the very high value that has accrued to individual personality through the conscience and dignity it is invested with; we know too how closely this regard for the individual is bound up with our whole social structure. Now it is inevitable that this sacred virtue which invests the individual should be extended to the things he is closely and lawfully connected with. The sentiments of respect for him cannot be limited to the physical person alone; the objects considered as his own must certainly have a share in them. Not only must this follow, but it serves a purpose too. Our moral structure implies a wide measure of initiative being left to the individual. For this initiative to be possible, there must be some kind of region where the individual is his own sole master, where he may act with entire independence, shield himself from any outside pressure and be truly himself. This individual liberty that we have so much at heart does not only assume the faculty of moving our limbs as we like; it implies the existence of a range of things that we can dispose of at will.

Individualism would be no more than a name if we had not some physical sphere of action within which we could exercise a kind of sovereignty. When we say that individual property is a sacred thing, we do no more than state in symbolic form a moral axiom that cannot be gainsaid, for the cult of the individual depends absolutely upon it.

But this is the characteristic of individual ownership rather than an explanation of it. What we have just set out allows us to understand how things *lawfully possessed* or owned are and must be invested with a characteristic that insulates them from being assailed; but the argument does nothing to tell us what qualifications the things must have before it can be claimed that they are lawfully possessed, that they legitimately form part of an individual domain. Not every thing that comes into relation with the individual, even into permanent relation, can be legitimately appropriated by him and so become his property. When, therefore, can appropriation be said to be rightfully founded on justice? This is something that could in no wise be determined by the sacred character that the person is invested with. In former times, when property was collective, the problem did not exist. For the right of ownership then had a quality *sui generis* as its source which was inherent in the things themselves and not in persons. There was no question as to what things the quality might be communicated to, since it resided in them. The whole problem was to know what persons could make use of this quality for their own benefit, and the answer went without saying. Those alone could make use of it—by the means we described—who knew how to turn it to proper account. But nowadays it is no longer the same. It is in the person that the attributes reside on which the ownership is founded. The question then arises—what connexion do the things have to maintain with the person for the sacred character of the person to be legitimately communicated to the things? For it is this communication that constitutes the taking of possession.

The only way to answer this is to study the various means of acquiring property and to attempt to isolate the principle or principles, and to see what foundation they have in our social structure. This acquisition has two main sources: that is, by contract and by inheritance. It would no doubt be simpler to confine ourselves to these two methods. There remain, apart from these, gifts or donations, and uninterrupted possession. The only gifts that play an important part in this sense are bequests by will or legacy, and since they are closely connected with inheritance, we shall be dealing with them at that point. As to uninterrupted possession, although it would make an interesting study as history, it does only account for a very small part in the various forms of property of the present day. The two main ways, then, in which we become owners, are exchange by contract, and inheritance. By the second, we acquire these properties ready made; by the first, we create new objects of ownership. It may be said that this would be attributing to the contract something that can only be the product of labour—since labour alone is creative. Labour in itself, however, consists exclusively in a certain expenditure of muscular energy:

it cannot, then, create things. Things can be only the recompense of labour; it cannot create them out of nothing; they are at the cost of labour just as they are the condition on which it is dependent. Labour can therefore bring property into existence only by way of exchange, and all exchange is a contract, explicit or implicit.

Now, of the two sources of property we have, one appears at first glance to be in conflict with the very principle on which property as we know it rests: I speak of individual property. In fact, individual property is of the kind that has its origin in the individual who owns and in him alone. The property which is the result of inheritance comes, by definition, from other individuals. It has been formed without his participation; it is not his creation; it can therefore only have a connexion with him that is quite exterior. We have seen that individual property is in conflict with collective property. Now, inheritance is a survival of the latter. When the family, formerly joint owners, breaks up, its original joint possession persists in another form. The rights that each member of the group had in the property of the others were as if immobilized and held in check during the lifetime of these others. Each one enjoyed his possessions separately; but from the time that the actual holder was about to die, the right of the former joint proprietors regained all its force and all its efficacy. In this way the right of succession came to be established. For a long period the right of family joint ownership was so strong and respected that, although the family no longer lived as a community, this right stood opposed to any actual holder being able to dispose of his property through donation, testamentary or otherwise. He had only a right of life interest: it was the family who was the owner. However, since owing to its break-up, it could no longer exercise this right collectively, it was the nearest relative of the deceased who took over his rights. Inheritance is therefore bound up with archaic concepts and practices that have no part in our present-day ethics. It is true that this fact alone does not warrant our thinking it is bound to disappear. We sometimes have to keep such survivals, where they are needed. The past persists beneath the present, even when they are at variance. Every social structure is full of these paradoxes. We can do nothing to cancel what has been—the past is a

174

reality and not to be done away with. The earliest forms of society have provided a foundation for the most recent: it often occurs that a continuity of some sort has been kept up whereby the older forms in part are preserved to nourish the newer. These points will be enough to show that of the two main processes by which property is acquired, inheritance is the one that is going to lose its importance more and more. All this persuades us that by analysing contractual rights we shall discover the principle upon which the institution of property is going to be founded in the future. We shall therefore set about this study.

CONCERNING THE CONTRACT

For most, the notion of the contract is of an operation so simple that it could be deemed to be the primary fact from which all other social facts have derived. It is upon this idea that the theory of the social contract rests. The outstanding example of the social bond which unites individuals within one and the same community must have been or ought to be one of contract. And if we thus take the contract as a phenomenon of early origin, considered either in time or, as by Rousseau, logically, this is because the concept of it seems clear in itself. It appears to have no need of being linked with some other concept that explains it. Jurists have often proceeded according to the same principle. In this way they have traced the origins of all obligations arising from man as a social fact, either to breaches of the law or to the contract. All other obligations that have not their source precisely in such breaches or in a contract, in the true meaning of the term, are held to be variations of these two. In this way the theory of the quasi-contract has been formed by which, for example, we explain the obligations arising out of the management of other people's affairs, or out of the fact that a creditor has received more than his due. The idea of the contract seemed so revealing in itself that the originating cause of these various obligations also seemed in itself to lose all obscurity as soon as it had been assimilated with the contract proper, that is, as soon as it had been set up as a kind of contract. Nothing, however, could be more deceptive than this seeming clearness. Far from the contract being of early origin as an institution, it does not appear, and above all, does not develop, until a very late date. Far from being simple, it is of extreme complexity and it is not easy to see how it took shape. It is important to understand this first of all. To be clear on this point, let us make a start by defining how the bond of contract itself is composed.

To begin with, what does a juridical moral bond consist in? This term may be applied to a relation conceived by the public consciousness as existing between two subjects, individual or collective, or between these subjects and a thing, and by virtue of which, one of the parties in question has at least a certain right over the other. As a rule, a right exists on both sides. But these mutual rights are not inevitable. The slave is bound in law to his master and yet has no right over him. Now, bonds of this kind may come from two different sources. (1) Either they derive from a state or condition in being, of things or of persons in relation; and it—the state or condition —may be such that these things or persons are (intermittently or permanently), of a certain nature, in some particular setting, and held by public consciousness to possess certain acquired characteristics. Or, (2) they derive from a state not yet in being of things or of persons, but simply desired or willed on both sides. In this case, then, it is not the intrinsic nature of the state or condition that brings about the right, but rather the fact that the state is willed or desired. The right consists in this instance simply in bringing about the state just as it has been willed. Thus, because I was born in a certain family and bear a certain name, I have duties towards certain persons who are related to me, or towards certain others over whom I may have to exercise some guardianship. Because a certain thing has effectually come into my patrimony by legitimate means, I possess rights of ownership over this thing. Because I own certain landed property having certain features of situation, I have a certain right of easement over the neighbouring estate, and so on. . . . In all these instances it is a fact, established or made a reality, that brings about the right I exercise. But I may come to an agreement with the owner of a house for him to lease his property to me, in return for a sum of money to be paid over to him every year on certain agreed terms. In this event there is simply the will on my part to occupy this property and to pay the sum promised, and on the part of the other the will to surrender his rights in consideration of the sum agreed upon. Here, however, there are only volitions, or states of will in question, and yet this state of wills may be enough to bring about obligations and therefore rights. It is for the bonds that have this origin that the term 'contractual' has to be reserved. Clearly, between the extremes of these two contrasting types, there are countless intermediary links. The main thing is to take a close look at these extremes so that we may the better appreciate the contrast. The complete divergence is indeed striking. On the one hand we have relations in due form according to law, having as their origin the status of persons or of things, or of the modifications so far latent in this status; on the other, relations according to law having as their origin wills that are in agreement to modify this status.

This definition shows at once that the bond of contract could not have had a very early origin. Indeed, men's wills can not agree to contract obligations if these obligations do not arise from a status in law already acquired, whether of things or of persons; it can only be a matter of modifying the status and of superimposing new relations on those already existing. The contract, then, is a source of variations which pre-supposes a primary basis in law, but one that has a different origin. The contract is the supreme instrument by which transfers of owner-ship are carried through. The contract itself cannot constitute the primary foundations on which the right of contract rests. It implies that two legal entities at least are already properly constituted and of due capacity, that they enter into relations and that these relations change their constitution; that some thing belonging to the one passes to the other and vice versa.

There may be, for instance, two families, A. and B. ; a woman leaves A. to go off with a man of B. and becomes in some respects an integral part of this group. A change has been brought about in the scale of the families. If this change occurs peacefully, with the consent of the two families concerned, we then get, in more or less rudimentary form, the marriage con-tract. Hence it follows that marriage, being of necessity a con-tract, pre-supposes an existing structure of the family that has nothing contractual about it. This is one proof the more that marriage rests on the family and not the family on marriage. If we can suppose that incest had not been always something prohibited, and that every man had been united to a woman of his own family, sexual union would not have involved any actual transfers either of persons or of things. The marriage contract would not have come into being.

We are agreed that the bond of contract is not of very early date. At the same time it is easy to see how it came to be devised only at a late stage. What indeed is the origin of bonds, that is to say, of rights and obligations that have their source in the state or condition of persons or of things? They derive in fact from the sacred character of one or the other, from the moral prestige they are endowed with, either direct or indirect. When early man considers himself under binding obligation to his own group, it is because this group is in his eyes the supremely holy thing. If he acknowledges his obligations equally towards the individuals forming the group, it is because something of the sanctity of the whole is communicated to its parts. All members of the same clan have within them, as it were, a particle of the divine being from which the clan is supposed to be descended. Thus, they bear the mark of a sacred symbol, and this is why they are bound to be defended, to have their death avenged, and so on. . . . We have seen too that the rights which have their origin in things, derive from the sacred nature of things; we need not revert to this. Therefore, all moral and juridical relations and ties which derive from a personal or from a *real* status, owe their existence to some virtue *sui generis*, inherent either in the subjects or the objects and compelling respect. But how could a virtue of this kind reside in mere inclinations of the will? What is there and what could there be about the action of willing some thing or willing some relation, which may compel this relation to be carried into practical effect?

First, let us have a look at this idea that the agreement of two wills bent upon a common aim may possess a character that makes it binding on each. Here we have an important innovation in law that argues a very advanced stage in its history. When I have resolved to act in a certain way I am always free to go back on my decision. Why then, should two decisions coming from two different individuals have a greater binding force, simply because they are at one in agreement? It is understandable that I should come to a halt in the presence of a being which I hold to be sacred, and that I should refrain from touching him or from modifying his status, whatever it be, owing to the qualities I attribute to him and to the respect he thus inspires in me. It is the same with things, which may be similarly placed. But an act of the will, a resolve, remains still only a possibility: this is, by definition, not a thing that has in any way materialized or become effective. How is it that a thing should be able to compel me to this degree, a thing which is not in being or which is in being only in the region of ideas? We may suppose that all kinds of factors must have intervened for our volitions to have become endowed with a compelling quality not found implicit in them when they are analysed. And so, the juridical notion of the contract, that is, of the contractual bond, far from being immediately self-evident, must have been built up by long endeavour.

Indeed, it was only by very slow degrees that societies succeeded in getting past the initial phase of the purely statutory right and in superimposing upon it a new right. It was only by successive changes in statutory right that they drew nearer by gradual stages to the new right. This development came about in various ways and the main ones are as follows.

New institutions begin as a rule by taking the old as their model and only split off from these by degrees in order to develop their own pattern in freedom. The function of the right of contract was to modify the status of the person; to bring this about, however, the right was conceived on the model of the statutory right. The bonds that unite persons by reason of their state, acquired and established, are dependent on this state. They come about owing to the fact of these persons having a quality in common that makes each respected by the other. To put it more precisely, members of the same clan or family have duties towards one another because they are supposed to be of the same flesh and blood. Not that this physical relationship in itself has any moral efficacy. But it means that the blood is the vehicle of a sacred principle which it mingles with: to be of the same blood is to share the same god and to have the same sacred quality. Very often, too, the ritual of adoption consists of introducing some drops of blood into the veins of the one adopted. It follows, then, that when men felt the need to create ties other than those of their own family status, that is, ties which they willed, they conceived them as a matter of course in the likeness of the only ties familiar to them. Two different individuals or groups, between whom no natural ties exist, agree to be associated in some common aim: in order that this covenant should be binding, they must bring about the physical blood relationship considered to be the source of all obligations. They mingle their blood. For instance, the two contracting may dip their hands in a jar into which a little of their blood has been made to drip, and they suck up a few drops.

It is this process that Robertson Smith studied under the name of 'blood covenant'; its nature and world-wide range are to-day well known. In this way, the two parties became under an obligation or pledge to one another; in some respects this relation was the result of an act of their will; there was an element of contract in it; but it acquired its whole efficacy only by taking the form of a relation by contract. The two individuals formed, as it were, a kind of artificial group which rested on ties similar to those of the natural groups to which each belonged. There were other ways that led to the same result. Food makes blood and blood makes life; to eat the same food means to commune together or partake of the same source of life; it means making the same blood. It is from this that we reach the important part played by the rite of communion by breaking bread in all religions, from the most ancient to Christianity. The same sacred thing is eaten in common, thus partaking of the same god. Those taking part became pledged by this act. The two persons contracting were equally able to bind themselves by drinking from the same cup, by eating from the same dish of food or even by sharing a meal. The act of drinking from the same cup is to be found in many wedding customs. The practice of sealing a contract by drinking together probably has the same origin, and likewise that of shaking hands.

In these examples we see bonds of the personal status serving as a pattern for the bonds of contract that were beginning to develop. But bonds of the *real* status were used for the same purpose. The rights and obligations that I have in regard to a thing are dependent on the state or condition of this thing, on its position under the law. If it forms part of the patrimony of another, I must respect it; if, nevertheless, it should come into my patrimony, I must restore it or hand over its equivalent. This being so, let us suppose there are two individuals or two groups wishing to make an exchange, for instance, to exchange one thing for another or for a sum of money. One of the parties hands over the thing; by this act alone, the one who receives it finds himself contracting an obligation, that of handing over its equivalent. Such is the origin of what has been termed the *real* contract, that is, a contract formed only by the actual delivery or handing over of a thing. We know the part played by real contract not only in Roman law and Germanic law, but in our own ancient French law. Very clear traces of it still remain even in the law of the present day. For it is from this source that the custom comes of giving earnest-money or *arles*. In place of giving the object of exchange itself, only part of its value was given or some other object. Very often a thing without value came to suffice, such as a straw, or the glove used in Germanic law. The object received made the one receiving it a debtor towards the other. In the course of time, it became enough to make a gesture of presenting the object. But as we see, neither the blood covenant nor the *real* contract was, properly speaking, a contract. For, in both cases, the obligation does not come about from the efficacy of the wills in agreement. These would, of themselves, have no binding force. They must over and above this cover a state or condition of persons or of things and it is in reality this state and not the contracting wills, that is the operative cause of the bond made. When, on the strength of a blood covenant, a man finds himself under obligations to his kindred by adoption and vice versa, that is not by virtue of the pledge given but because the covenant has this force of making them of the same blood. If, in a *real* contract, I am in debt for the price of the object received, it is not because I have promised it but because this object has passed into my patrimony, because it has henceforth a certain position under the law. All these practices are just so many methods of reaching almost the same result as a contract made by means other than the contract proper. For, we must repeat, what constitutes the contract is the declared agreement of the wills concerned.

And there must be something more than this: a state or condition of things or of persons without any intermediary must have been set up, of a kind to bring about effects in law. As long as an intermediary exists, there can be no true contract.

There is, however, another way of getting nearer to the contract proper. The wills can effect a bond only on condition of declaring themselves. This declaration is made by words. There is something in words that is real, natural and living and they can be endowed with a sacred force, thanks to which they compel and bind those who pronounce them. It is enough for them to be pronounced in ritual form and in ritual conditions. They take on a sacred quality by that very act. One means of giving them this sacred character is the oath, or invocation of a divine being. Through this invocation, the divine being becomes the guarantor of the promise exchanged. Thereby the promise, as soon as exchanged in this way (and even though it should not reach the point of execution), becomes compulsive, under threat of sacred penalties of known gravity. For instance, each contracting party pronounces some phrase that binds him and a formula by which he calls down upon his head certain divine curses if he should fail in his undertaking. Very often, sacrifices and magical rites of all kinds reinforce still further the coercive force of the words uttered.

This, then, seems to be the origin of contracts made in all due and solemn formality. One of their features is that they are binding only if the parties make an undertaking by a formula, solemn and agreed, which cannot be evaded. It is the formula that binds. This is the distinctive sign by which we recognize a main feature of magic and sacred formulas. The juridical formula is only a substitute for sacred formalities and rites. When certain definite words, arranged in a definite sequence, possess a moral influence which is lost if they are different or merely pronounced in a different sequence, we can be certain that they possess or have possessed a sacred significance and that they derive their peculiar powers from sacred causes. For it is only the sacred phrase which has this effect upon things and upon human beings. With the Romans especially, one fact tends to show clearly that the origin of the contract had a sacred character: this is the custom of the *sacramentum*. When two contracting parties were in disagreement on the nature of their respective rights and duties, they deposited a sum of money in a temple, which varied according to the importance of the dispute: this was the *sacramentum*. The one who lost his case also forfeited the sum he had deposited. This means he was fined to the benefit of the deity, which argues that his project was held to be an offence against the gods. These gods were, then, party to the contract.

We can now see how slow the notion of the contract was in developing. The blood covenant and the *real* contract are not true contracts. The solemn agreement or contract is closer to it. For in this case the undertaking is sacred as soon as the wills in question have been declared by words accompanied by consecrated formulas. Even in this instance, the moral value of the undertaking does not come about through the consent of the wills but by the formula used. If the solemn ritual is lacking, there is no contract. In the next lecture we shall see the stages the right of contract had to pass before reaching its present position.

XVI

MORALS OF CONTRACTUAL RELATIONS
(Continued)

WE saw in the last lecture how difficult societies found it to rise to the concept of an agreement or contract. All rights and duties derive from a state that has come into being of things or of persons; but in the contract proper, it is a state simply conceived and not yet in being that lies at the root of the obligation. Nothing has been acquired or given except a declaration of will. How is it possible for such a declaration to bind the will from which it springs? Should we say that in the contract there are two wills concerned and that they bind each other mutually in some way; that they have become bound up together in some fashion and that this association does not leave them wholly free? But how can the promise made by the other contracting party to fulfil certain terms of performance—if on my part I fulfil certain other terms of performance—compel me to honour my promise and vice versa? It is not because the other has pledged himself towards me that my undertaking towards him is binding in greater or lesser degree. The one undertaking is not of a different nature from the other; and if neither has in itself the moral authority to compel the will, it will not be got by their agreeing. Moreover, in order that there shall be a contract, there is no need for an undertaking of reciprocal performance. There may also be unilateral contracts. For instance, deeds of gift and contracts under guarantee do not involve any exchange. If, in a case of this kind, I declare that I will give a certain sum or some object to a certain given person, I am bound to carry out my promise although I have received nothing in exchange. Therefore in this case it is solely the declaration of my will, without any reciprocal declaration, that binds me. How does it come to have this particular force?

It was only by very slow degrees that the people of various nations reached the point of endowing this demonstration of will alone with such force, both moral and juridical. When exchanges or transfers of property became more frequent and a need for contractual relations began to make itself felt, expedients were arrived at to meet this need. Without founding a new right, an attempt was made to adapt the statutory right to these new demands. This was the principle adopted. As soon as the parties were in agreement, a state or condition of things or of persons was brought into being, which carried further obligations in its train. For instance, one of the contracting parties would carry out the terms of the performance he had undertaken. Henceforward, there was something acquired and established that bound the other party. The vendor delivered the thing; the thing, which in this case passed into the patrimony of the purchaser, put an obligation upon him to fulfil his part: this was by virtue of the principle—everywhere accepted but differently observed by various societies—that no one may enrich himself at the expense of another. Or, it might be that once the terms of agreement were settled, the contracting parties submitted themselves to a process that created a kind of kinship *sui generis* between them; this acquired kinship then introduced between the two a whole system of rights and duties.

By these two processes, a change was introduced in the statutory right in consequence of an agreement of wills, and the bonds made come to have a contractual character. But these bonds are not the product of the agreement of wills, and in this respect there is still no true contract. In the two cases, the consent of itself has no power to compel; at least, it produces no rights, except through an intermediary. It is a state, acquired by things or by persons, following upon the understanding but without any intermediary, that causes the understanding to have juridical effects. As long as the performance has not been carried out, at least in part, and as long as the contracting parties have not mingled their blood or have not sat at the same table, they remain free to revoke their decision. Thus, the declaration of the will by itself alone is devoid of any efficacy. The statutory right has been used to achieve almost the effects of the contractual right but this right has so far still not come into existence.

There is, however, another way in which men succeeded in getting nearer to it. Whatever the process, the wills can bind themselves only on condition of declaring themselves exteriorly, of projecting themselves outwardly. They have to become known, so that the society may attach a moral significance to them. This declaration or outward manifestation is done by the aid of words. Now, there is something in words that is real, natural and living, something that can be endowed with a sacred force thinks to which, once pronounced, they should have the power to bind and compel those who pronounce them. It is enough for them to be uttered according to a certain ritual form and in certain ritual conditions. Henceforward, they become sacred. We can well imagine that words, once they have assumed this sacred quality, impose respect on those who have uttered them. They carry the same prestige as those persons and things which are themselves the object of rights and duties. They too, therefore, may be a source of obligations. One means of conferring this quality on them and hence this binding force, is the oath or invocation of a divine being. By this invocation, this being becomes the guarantor of the promises made or exchanged: he is present and communicates to them something of himself and of the sentiments he inspires. To fail in this is to offend, to be exposed to his vengeance, that is, to sacred penalties which seem to the believers as certain and inevitable as those later to be imposed by the tribunals. In these circumstances, once the words have left the lips of the contracting party, they are no longer his own, they have become something exterior to him, for they have changed in nature. They are sacred and he is profane. In consequence, they are beyond his free-will: although they come from him, they are no longer dependent on him. He can no longer change them and he is bound to carry them out. The oath, too, is a means of communicating to words, that is, to the direct manifestations of the human will, the kind of transcendence we see in all moral things. Also, it separates the words, as it were, from the person who utters them and makes something new of them, which is then imposed on the speaker.

Such, then, certainly seems to be the origin of contracts made in all due and solemn formality. One of their features is that they are not valid unless certain agreed formulas have been pronounced. This cannot be evaded: otherwise the contract has no binding force. An essential feature of magic and sacred formulas can be recognized by this sign. When certain definite words, arranged in a given sequence, are held to have a force which is lost if the slightest change is introduced, we can be sure that they have or have had a sacred meaning and that they derive their peculiar powers from a sacred source. For it is only the sacred phrase that has this effect upon men and things. Juridical formalism is only a substitute for sacred formalities and rites.

As far as the Germanic tribes were concerned, the word used to describe the act of making a solemn contract is *adhramire* or *arramire* which is translated by: *fidem jurejurendo facere.* Elsewhere we find it combined with the *sacramentum:* "*Sacramenta quae ad palatium fuerunt adramita.*" *Adhramire* is to make a solemn promise on oath. It is extremely likely that in the beginning the Roman *stipulatio* had the same character. It is a contract made *verbis*, that is, by established formulas. Anyone knowing to what degree Roman law was in principle a sacred and pontifical thing, will have little doubt that these *verba* were at first ritual formulas intended to give a sacred character to the undertaking. It is certain they were pronounced in the presence of priests and possibly on sacred ground. Were not solemn words, moreover, called sacramental words?

It is probable that very often, if not always, the verbal ritual did not suffice to sanctify the words exchanged and make them irrevocable; manual rites were also employed. Such was the origin most likely of the *denier à Dieu* (the Lord's pence or gratuity). This was a coin that one of those contracting gave to the other once the bargain was clinched. It was not an instalment, or earnest-money, on the price, a kind of *arles* or pledge, for it was an extra payment provided by one of the parties which was not charged to the sum finally paid. It does not therefore seem possible to read into it a fulfilment in part such as we see in *real* contracts. It must have a meaning. As a rule, it was used in religious practices and this is shown in its name : *denier à Dieu*. Might it not be a survival, rather, of some offering intended to interest the deity in the contract in some way, to make him party to the arrangement? This seems quite as effective a way of invoking him as the spoken word and thus, of giving sanction to the undertaking proposed.

The same probably applies to the ritual attached to the straw stalk. In the last lecture we seemed to see in this rite a survival of the *real* contract. We were mistaken, however. There is in fact no reason to believe that it is any less ancient than the *real* contract and hence there is no evidence to show it derives from that particular form. Another fact against associating the two is that the straw stalk or *festuca*, the delivery of which sanctified the bond contracted, was handed over, not by the future creditor, but by the future debtor. It was not, then, like the handing over in the *real* contract, a performance completed, in full or in part, since the performance on the part of the debtor still remained to be carried out in full. Such a procedure could not have the effect of binding the creditor to the debtor but rather of binding the debtor to the creditor. Finally, the solemn contract of the Romans that was made *verbis*, that is, by means of hallowed formulas, bore the name *stipulatio*. Now the word *stipulatio* comes from *stipula*, also meaning the straw stalk. And " *Veteres, quando sibi aliquid promettebant, stipulam tenentes frangebant.*" The stipula remained in popular use up to fairly recent times. This means that it had a close connexion with the solemn verbal contract. The two procedures seem inseparable. It is difficult to tell the exact meaning of this ritual. Evidently it signified a kind of liege homage or tribute by the debtor to the creditor, binding on the debtor. This made something of the legal identity of the debtor pass to the creditor, that is, some part of his rights. What makes me think that this is the true explanation is the nature of the procedure that took its place in the Middle Ages that succeeded that early period. The *festuca* indeed hardly survived the Frankish epoch. A gesture of the hand took its place. When it was a matter of an undertaking that had to be given towards some particular person, the future debtor placed his hands in those of the creditor. Where a unilateral promise alone was concerned or the affirmation of an oath (an affidavit), the hand was placed on holy relics or raised (was it to call heaven as witness?). We can appreciate the sacred, not to say the mystic character of these gestures, since they are still sometimes used, even in our own day; at any rate there is no doubt that their purpose was to create a bond. We are especially aware of this in two kinds of contract of primary importance. First, the feudal contract that bound the man to his lord. To plight his troth and homage, the man knelt and put his hands in those of his lord, promising fealty. The same practice is to be found in the contract of betrothal. Those betrothed engaged to marry by joining their hands and the ritual of Catholic marriage still has the traces of this. We know too that the contract of betrothal was binding.

We are no longer able to say precisely what the religious beliefs were that lay at the root of these practices. Some general indications do, however, come out of these comparisons. The laying on or joining of hands is a substitute for the handing over of the *festuca*; both therefore have the same meaning and the same purpose. The laying on of hands at any rate is very well known. It has been used in all religions. When it is a question of blessing or consecrating any object, the priest places his hands upon it. When it is a matter of the individual shedding his sins, he places his hand on the victims he is about to sacrifice. Whatever there is base or sinful in him and in his personality leaves him and is communicated to the beast and destroyed with it. It is by a process of the same sort, in another instance, that the victim offered up to do homage to some deity, becomes the substitute for the person who sacrifices it or has it sacrificed. Thus, the personality— either as a whole or in some particular part—appeared to men's minds to have the quality of something communicated or communicable. And clearly, the function of these rituals was to bring about communications of this kind. It is true that when we examine them against our ideas of the present day, we are inclined to see only symbols, and ways of giving allegorical form to the bonds contracted. But as a rule customs never assume a symbolic character in the beginning; the symbolism represents only a decadence that comes when the primary meaning of the custom is lost. Customs begin by being active causes, and not symbols, of social relations; they bring these relations into existence and it is not until later that they decline into being mere external and material indications. The transfer lying at the base of the *real* contract is known to be a real transfer and it is this that makes the contract and gives it its binding force. It is much later that it becomes simply a means of giving material proof of the existence of the contract. The same thing applies to the customs we have just examined. We are justified in relating these to the blood covenant. They, also, have the effect of binding the contracting parties by taking on their moral personality. It may be that the hand-shake (*Handschlag*) has the same origin.

Contracts of this kind, then, are made up of two elements: a verbal nucleus, which is the formula, and the outward ritual. As such, they are already nearer to the true contract than to the *real* contract. True, there must still be intermediary processes for the consent to have effects in law, but even so the wills concerned are bound direct by these very processes. In fact, these intercalated processes do not consist of actual performance, even in part, of what the contract sets out to do. Whatever the solemn ritual employed, the undertakings entered into by the two parties remain to be carried out in full, even after the rites have been observed. On both sides there are only promises and yet these promises commit the two contracting parties. This is not so with the *real* contract, since one of the two parties has already carried out his promise wholly or in part—that is, one of the two wills concerned is no longer in a state of willing, since it has reached its object. It is true that the blood covenant had the same advantage. It is easy to see, however, that this unusually complex ritual could only serve for great occasions and not for the small affairs of everyday life. It could not be used to ratify day-to-day sales and purchases. It was hardly ever used except to create some permanent association.

The contract with solemn ritual lent itself easily to the advance that was to come about in the course of time. The outward ritual forming its cloak, as it were, tended by degrees to wear thin and disappear. In Rome, these advances came about as early as the classical era. The outward formalities of the *stipulatio* had become no more than an echo from times past; only scholars find traces of them in customs of the people and current traditions, or in the derivation of the word. But they were no longer indispensable for the *stipulatio* to be valid. It consisted exclusively of the sanctified formula that the two contracting parties had to pronounce with religious punctilio. The same phenomenon appeared in modern societies under the influence of Christianity. The Church tended more and more to make the oath the necessary and adequate condition of a contract, without further formality. Thus, the intermediary process between the agreement of the wills and the obligation to translate this agreement into fact went on shrinking. Since words are the direct expression of the will, there remained—as conditions exterior to the consent itself—no more than the clearly defined character of the formula in which this consent had to be expressed, and the particular force and qualities attaching to this formula. When this force reached vanishing point, and as a result there was no longer any insistent require-ment as to the verbal form used by those contracting, then the contract proper, the consensual contract had arrived.

That is the fourth stage of this development. How then was the point reached when the contract was rid of this last ex-traneous and casual element? Several factors contributed to this result.

First, when exchanges in trade greatly increased in number and kind, it became difficult to keep up the practice of the solemn contract with its hampering formalities. New trading relations were set up by means of contract, which the stereotype formulas hallowed by tradition did not fit in with. The processes of law themselves had to become more flexible to conform to the social life. When sales and purchases were going on all the time, when there was never an instant that trade was at a standstill, it was not feasible to require every buyer and every seller to take an oath or have recourse to certain formulas laid down, and so on. . . . The day-to-day character, and continuity of these relations inevitably excluded all solemn ritual, and it was quite natural to seek means to lessen or lighten the formal-ities and even to do away with them. But this explanation is not enough. Because these means were needed, it did not follow that they were to be found. Let us see how they appeared to the public mind at the time they were found to be necessary. The mere fact that an institution is required does not mean it will appear at a given moment out of the void. There must be something to make it of, that is, current ideas must allow it to come about and existing institutions must not oppose it but, rather, supply the material needed to shape it. So it was not enough for the consensual contract to be demanded by the advance of economic life: the public mind, too, had to be ready to conceive it as possible. Until then, it had seemed that con-tractual obligations could only come about by prescribed ritual or by the actual transfer of the thing. Now, a change had to be made in the region of ideas, that would allow of their having a different start. That is how the last stage in these changes was brought about.

What was it that from the outset ran counter to the notion of the consensual contract? It was the principle that any obligation under the law could only have its origin, it seemed, in a state in being of things or of persons. In itself, this principle is irrefutable. Every right has a *raison d'être* and this can only lie in some clearly defined thing, that is, in an established fact. But is it impossible for mere declarations of will to satisfy this requirement? In no wise. It is true they cannot fulfil this condition if the will that has been asserted remains free to retract. For then the will would not amount to an established fact, since it could not be known in which direction it would finally manifest itself; nor could we say with any assurance what it was or what it might become. Therefore, nothing definite could result from it and no right could derive from it. But let us suppose that the will of the contracting party is asserted in such a way that it could not retract. In that case it will possess all the characteristics of the established fact, a fact in being and able to bring about consequences of the same kind, since it is irrevocable. I may engage to sell or lend to you a certain object, and on such terms that once the undertaking is made I have no longer the right or means to break it. If I do this, I arouse in you by this very action a state of mind equally decided and in line with the certainty you are justified in assuming about my action. You count, and legitimately so, on the promised performance. You have a right to consider it as about to take place, and you act or may act in consequence of this. You may make a certain decision or decide on a certain sale or purchase, by reason of this legitimate certainty. If I then suddenly withdraw and deprive you of this certainty, I throw you out just as seriously as if I had withdrawn after delivery of the thing I had assigned to you in making a *real* contract with you; I bring about a change in your established position and I render any transactions you may have engaged in on the good faith of the given word, ineffectual. We begin to see, too, morals ranged against this unjustified wrong.

Now, in the solemn contract, the condition we have discussed is fulfilled; the irrevocability of the will is made certain. It is the solemn ritual nature of the undertaking that gives it this characteristic, by sanctifying it and by making of it something that no longer depends on myself, although proceeding from me. The other party is thus justified in counting on my word—and vice versa, if the contract imposes mutual obligations. He has morally and legally the right to consider the promise as inevitably about to be kept. If, then, I fail in this, I am transgressing two duties at once: (1) I am committing sacrilege, because I am breaking an oath, I am profaning a sacred thing, I am committing an act forbidden by religion, and I am trespassing on the region of sacred things. (2) I am disturbing another in his possession, just as if I were a neighbour on his land; I am injuring him, or there is a danger of it. From the very moment the right of the individual is properly respected, that individual shall not be done any unmerited wrong. Thus, in a solemn contract, the formal bond that ties the contracting parties is a twofold one: I am bound by my oath to the deities; I have an obligation to them to fulfil my promise. But I am also bound towards a fellow-man, because my oath, by detaching my word and exteriorising it, enables this fellow-man to possess himself of it as of a thing. There is, then, a twofold resistance to such contracts being broken, partly in the ancient and sacred rights and partly in modern and human rights.

We can now get some idea of how these things came about. It is the second of these elements which, detached from the first and entirely rid of it (e.g. of the solemn formalities), has become the consensual contract. The demands of a busier life tended to reduce the importance of ritual formalities. At the same time, however, the decline in faith lessened the value attached to them and by degrees, the meaning of many became lost. Thus, if there had been in the solemn contract no more then the legal ties originating in solemn ritual, this development would have ended in a truly backward step in contractual rights, since the undertakings contracted would henceforward have lacked any foundation. But we have just seen that there was another form of contract which managed to survive: this was the contract that has its roots in the right of the individual. It is true that this second type of bond is an offshoot of the first; for if there is already an established fact, if the spoken word assumes an objective character that removes it from the disposal of the contracting party, it is because an oath has been taken. But could this result, which used to be got in this way, now be reached by other means? It is enough to establish that the declaration of will alone was irrevocable—that is, if it was made without reservations or concealment or hypothetical conditions, if, in a word, it was represented to be irrevocable; from then onwards it could in this way have the same effect with regard to individuals as when it was hedged about with solemn formalities, and it had an equally binding force. That is to say, the consensual contract had come into existence; we must therefore derive it from the contract by solemn ritual. Whilst this ritual form of contract had taught men that undertakings could be made by a clearly defined procedure, it is true to say that this clearly defined quality came from ritual processes and liturgy. This quality was detached from the cause that originally produced it and linked to another cause—hence a new form of contract came into being, or rather, the contract proper. The consensual contract is a contract by solemn ritual—of which the useful effects are preserved, although they are reached by a different procedure. Had it not been for the existence of the contract by solemn ritual, there would have been no notion of the contract by mutual consent. Nor would there have been any idea that the word of honour, which is fugitive and can be revoked by any-one, could be thus secured and given substance. The contract by ritual was secured only by magic and sacred processes: in the consensual form the given word acquired the same security and the same objectivity through the effect of the law alone. If we are to understand this new form of contract, we cannot proceed from the nature of the will or the words that declare it: there is nothing in the word to bind the individual pronouncing it. The binding force, the action, are supplied from without. It is religious beliefs that brought about the synthesis; once formed, other causes sustained it, because it served a purpose.

Naturally, this is a simplification, to make these matters more intelligible. The system of formalism was not done away with from one day to the next and the new principle established. It was only by very slow degrees that the solemn ceremonies lost ground, under the dual influence we mentioned: that is, the new demands of economic life and the gradual clouding over of the concepts that lay at the root of these solemn rites. It was, too, only very gradually that the new rule shed the formalistic wrappings that enclosed it. This happened only as the need for it became more urgent and when the old traditions weakened as a counter-force. The conflict between the two principles lasted a very long time. Both the *real* contract and the contract by ritual remained the basis of the Roman contractual right, which was preserved to apply only in certain cases. And clearly, there are very distinct traces of the early juridical concepts to be found well into the Middle Ages.

The contract by solemn ritual has moreover not wholly passed away. In all codes of law it still has some application. This discussion will have enabled us to understand what these survivals amount to. The contract by solemn ritual binds men doubly; it binds them one to the other; it also binds them to the deity, if it is the deity that was party to the contract; or to the society if the society took part in the person of its representative. Further, we know that the divinity is only the symbolic form of the society. The contract in solemn ritual form therefore binds us more strongly than any other. This is why we are compelled to its use whenever the bonds to be forged are of supreme importance, as in marriage. Now, marriage is a contract by solemn ritual not just because the religious ceremonies provide the evidence and record the dates, etc. . . . . It is above all because the bonds that have called forth moral values of a high order cannot then be broken at the arbitrary will of the parties. The intention is that a moral authority that stands higher should be mingled with the relationship being formed.

XVII

THE RIGHT OF CONTRACT (End)

THE consensual contract (or contract by mutual consent), in the final analysis, is, as it were, a climax, a point of convergence where the *real* contract and the ritual verbal contract meet in their process of development. In the *real* contract there is the transfer of a thing and it is this transfer that gives rise to the obligation; in receiving a certain object that you hand over to me, I become your debtor. In the contract by ritual, no performance takes place; everything is done by words, usually accompanied by certain ritual gestures. These words, however, are pronounced in such a way that they have hardly left the lips of the promisor when they become, as it were, exterior to him; they are *ipso facto* withdrawn from his option; he can have no more effect on them; they are what they are and he can no longer change them. They have thus become a thing, in the true meaning of the word. But then they too become transferable; they too can be alienated in some way or, like the material things that make up our patrimony, transferred to another. Those expressions still in current use—to give one's word, to pledge one's word—are not mere metaphor: they correspond in fact to a parting with something, a genuine alienation. Our word, once given, is no longer our own. In the solemn agreement or contract by ritual, this transfer had already been achieved, but it was subject to the magic-religious processes we have mentioned, which alone made the transfer possible since it was these ceremonies that gave an objective character to the word and to the resolve of the promisor. Once this transfer sheds the ritual that was previously a condition of it and is rid of it, once it constitutes the whole contractual act in itself alone, then the consensual contract has come into existence. Now, given the contract by ritual as an existing fact, this process of cutting down and simplifying was bound to come about of itself. On the one hand we see a diminishing of the verbal or other ritual ceremonies was brought about by a kind of decline from within, under the pressure of social needs that called for greater speed in the process of exchanges. On the other hand, the practical effects of the contract by ritual could be got adequately by means other than these ceremonies; it was enough for the law to declare as irrevocable any declaration of the will presented as such: this simplification was the more easily allowed of, since in the natural course of time the practices that had accumulated had lost a great part of their meaning and early authority. It is true that the contract so reduced could not have the same binding force as the contract by ritual since in the latter the individuals are, as it were, doubly bound—bound to the moral authority that intervenes in the contract, and bound one to the other. But it was at that very time that economic life needed some loosening process to slacken the stiff knots of contractual ties; if they were to be made with greater ease, it was imperative they should have a more secular character, and the act designed for making the ties had to be freed of all traces of ritual solemnity. It was enough to reserve the solemn contract for cases where the contractual relation had a special importance.

Such then, is the principle of the consensual contract. To sum up finally, it consists in substituting for the material transfer by *real* contract a transfer that is simply oral and even, to be more exact, mental and psychological, as we shall see. Once it was established it entirely took the place of the *real* contract, which henceforth had no more *raison d'être*. The binding force of the real contract was no greater, and furthermore, its forms were unnecessarily complex and general in terms. This is why it has left no traces in our present laws, whilst the ritual contract survives alongside the consensual contract which grew out of it.

As this principle becomes established, it brings about various changes in the contract as an institution, and these, little by little, will alter its whole aspect.

The system of the real contract and the ritual contract correspond to a stage in social evolution in which the right of individuals commands only a slight measure of respect. The consequence was that the individual rights concerned in every

contract had only a very slight measure of protection. It is true
the defaulting debtor often was sentenced to a penalty such as
whipping, imprisonment or a fine. In China, for instance, he
receives a certain number of strokes with a bamboo cane; in
Japan, there is the same custom; according to ancient Hindu
law, the penalty is a fine. But the rule was then still unknown,
whereby the true sanction is one of compelling the contracting
party either to keep his word, or to make good the loss or injury
inflicted on the other party by failing to honour an undertaking.
In other words, penal sanctions at this stage are only applied in
respect of a contract where it appears to be an offence against
the public authority: the way in which it affects the individual
does not enter into consideration. For loss or injury in private
cases there is no provision.

The result was that the creditor had no assurance whatever
of getting his debt paid. It is no doubt this situation that is responsible for a curious custom to be seen in various countries,
but especially in India and Ireland: it is generally known by the
name given it in India, the '*dharna*'. In order to enforce the
settlement of the debt, the creditor instals himself at the door
of the debtor and threatens to let himself die of hunger there,
if he does not get satisfaction. If the threat is to be taken as
serious, the man who is fasting must of course be resolved on
carrying it out to the end, if needs be, even to the point of
suicide. We are told by Marion, in listing the legal means to
compel a debtor, that . . . "In the fourth place, there is the
process of fasting, when the creditor places himself at the door
of the debtor and there lets himself die of hunger ". . . . The
efficacy of the strange proceeding derives from the beliefs and
sentiments connected with the dead. We know they were held
in the greatest awe. They are the powers from whom the living
cannot escape. It often happens, too, in the lower societies,
that a suicide can be in the nature of a vendetta. It is believed
that a man is avenged on his enemy with greater certainty by
killing himself than by killing him. Above all, it is a method of
revenge that the weak can use against the strong. A man may
be able to do nothing, in this life, against some powerful
character. He always has the power, in place of the earthly
revenge that he cannot inflict on his enemy, to take a revenge
from beyond the grave that is held to be more terrible and

above all, infallible. It might even be quite possible, in the case of the *dharna* in its true sense, for the suicide to have a twofold aim—that of immuring the debtor in his house whilst giving a magic potency to the threshold that would render it impassable. It is indeed on the threshold that the creditor instals himself and there that he dies; so it is to this spot that his spirit, once released from the body, will return. The spirit will keep guard over the threshold and will hinder the actual owner from crossing it. At least, he will cross it only at his peril. So that it is like a kind of mortmain put on the house, a kind of post-humous distraint.

Such a custom obviously shows that the creditor is left to his own devices to obtain satisfaction. Moreover, even in Germanic law, it is he who has to carry out the distraint. True, the law requires the debtor to allow it to be done, but the authority does not intervene in behalf of individuals and does not even assist them. This is because the specific bond contained in the contract did not have a very definitely moral character: it took on this character only in the consensual contract, for in this it forms the sum total of the relationship. Thus, the penal sanctions applying to the contract consist in essence not in requiting the public authority for some breach of the law, as by the defaulting debtor, but in ensuring for both parties the full and direct carrying into effect of the rights they had acquired.

It is not, however, only the sanctions—that is, the outer structure of the contractual right—that have been modified. The internal structure was entirely transformed.

In the beginning, the formal or ritual contract, like the *real* contract, could only be unilateral. In the *real* contract, the unilateral character arose by the fact that one of the parties carried out a performance indirectly; he could therefore not be bound towards the other. There was only one debtor (the one who had received), and one creditor (the one who had delivered the thing). In the ritual contract it amounted to the same thing, for this form involves an individual who promises and one who receives the promise. In Rome, for example, it would be asked: "Dost thou promise to do or to give this or that?" The other would reply: "I do so promise.' To create a bilateral bond, that is, in order that there shall be an exchange in the course of the contract, in order that each contracting party shall be both debtor and creditor at once, there had to be two separate contracts, independent one of the other, for the role assigned to each was entirely different. There was of necessity an actual transposition. The one who spoke first as stipulator or creditor later on spoke as debtor and promisor, and vice versa. The independence of the two processes was such that the validity of the one was entirely distinct from that of the other. Let us suppose, for example, that I have solemnly engaged to pay a certain sum to Primius as the price of a murder that he on his part has engaged to commit: this reciprocal obligation will take shape under the terms of a ritual contract, on the strength of two unilateral contracts in succession. I shall begin by solemnly promising a sum of money to Primius, who will accept; here, I am the promisor and he the stipulator and there is no question of a murder to be carried out. Then, by another contract, he will promise to perpetrate the murder at my request. The second contract is unlawful, because the cause of it is immoral. But the first is quite lawful: therefore Roman law would consider the promise to pay the sum of money as valid in itself and there would have to be recourse to a legal subterfuge to escape the consequences.

A system of this kind could therefore not lend itself easily to questions of exchange or to reciprocal or bilateral relations. In fact, in Germanic law, bilateral contracts were not unknown, but they appear only as transactions on a cash basis and such a transaction is not a truly contractual one. The consensual contract alone was able at a single stroke to create the two-way track of bonds that we find in any reciprocal agreement. For the greater flexibility of the system allows any contracting party to play at one and the same time the dual role of debtor and creditor, of stipulator and promisor. As a man is no longer under compulsion to adhere strictly to a definite formula, the reciprocal obligations can be contracted simultaneously. The two parties declare at one and the same time that they consent to the exchange on the conditions agreed between them.

Another new feature of some significance arose when consensual contracts became inevitably contracts in good faith (or *bona fide* contracts). This name is given to contracts whose range and legal effects must be exclusively determined by the intent of the parties.

The *real* contract and the consensual contract were not able to claim this characteristic, or at least, only very imperfectly. Indeed, in each case, the obligation did not come about purely and simply from the consent given or from the demonstration of will. Another factor needed to bind the parties came into it. Therefore this very factor which was indeed the decisive one, was bound deeply to affect the nature of the form of both these contracts; so it was impossible that these two forms of contract should depend exclusively or even mainly on what we might call the psychological factor, that is, the will or intention. In the case of the *real* contract there was the thing, the subject of the transfer; since the binding force of the act came from this thing, it contributed to a large extent to determining the scope of the obligation. In the Roman *mutuum*, a simple loan, the borrower was liable to repay things of the same quality and quantity as those he had received. In other words, it is the nature, kind or quantity of the things received that determine the nature, kind or quantity of the things to be repaid. This, then, is the original form of the *real* contract. Later on, it is true, the *real* contract served in exchanges (in the true meaning) in which the debtor owed, not a thing equivalent to what he had received, but an equivalent value. Here, the thing played a smaller part. But the use of the *real* contract for this purpose came relatively late; when it does take this form, it means that the consensual contract is in sight. Furthermore, as we said about the Germanic law, until it did appear, exchanges were hardly ever made except as a transaction on a cash basis. And so, even in this case, the thing delivered is no less a source of the obligation and thus, has a bearing on this obligation. It is not a matter of wondering what one of the parties had meant to deliver or what the other had meant to receive, since the delivery has been made and the thing is there, with its intrinsic value determining the value that the debtor owes to the creditor. The object speaks for itself and it is the object that decides. The role played by the thing in the *real* contract is filled by the words or the ritual used in the formal contract. Here it is the words used and the gestures that make the obligation; it is these, too, that define it. In order to know what the promisor or debtor is bound to give or to do, we must not consider his intent or that of the opposite party, but the formula he has used. The legal analysis, at least, has to start with the formula. Since it is the words that effect the binding, it is the words, too, that give the measure of the bonds forged. Moreover, even in the worst days of Roman law, the contract by stipulation had to be strictly interpreted. The intent of the parties, it is obvious, remained without effect whenever it was not possible to make it derive from the words used (Accarias, *Précis de droit romain*). For, we must repeat, the formula has a value in itself, that is, has its own force, and this force could not depend on the wills of the contracting parties since, on the contrary, the formula imposes itself on these wills. This is how a magic formula produces its effects, mechanically, as it were, no matter what the intentions of those using it. If these individuals know the way of using it best suited to their interests, all the better for them. But its action is not subject to their desires. For all these reasons, good faith and the intent of the parties hardly came into the reckoning, whether for *real* or ritual contracts. In Rome, it was only in the year 688 A.U.C. that the action known as *dolo* was instituted, allowing the contracting party deceived by fraudulent intrigue, to get reparation for the loss or injury caused.

From the time that the consensual contract was established, however, it was a different matter. Here we no longer find anything intervening in the relation contracted and affecting its nature. Certainly there are still words being used, as a rule at least, but these no longer have any force in themselves because they lack any sacred character. Their only value now is in giving expression to the wills they reveal and therefore in the end it is the state of these wills that decides the obligations contracted. The words in themselves are no longer of importance; they are only symbols to be interpreted, and what they signify is the state of mind and will that inspired them. We said just now that the expression 'to give one's word' was not altogether metaphorical. There is indeed some thing that we give, that we part with, that we are prohibited from changing. But strictly speaking, it is not the words pronounced which are marked *ne varietur*; it is the resolve they interpret. What I am giving to another is my firm resolve to act in a specified way: therefore

it is this resolve we have to get at to know what I have given,
that is, what I have pledged myself to do. For the same reason,
if a contract is to be achieved, the main thing is that it shall
exist in the intention or will of the party to it. If the will is lacking on either side, there can be no contract. For what the one
in fact is giving, is his intention of acting in a certain way, of
transferring his ownership in a certain object; what the other
declares, is his intention of accepting what is thus transferred
to him. If the intention is absent, nothing remains but a form of
contract empty of any positive content. All that is pronounced
is words devoid of meaning and so, devoid of value. We do
not have to specify the rules by which the intent of the parties
has to be appraised in its influence on contractual obligations.
It is enough to state here the general principle and to shew
how the consensual contract had to be a *bona fide* contract,
and how it could not be one of good faith except on condition
of its being one by mutual consent.

We can see how far the consensual contract amounts to a
revolutionary innovation in the law. The dominant part played
in it by consent and the declaration of will had the effect of
transforming the institution. It differs from the earlier forms of
contract from which it descends, by a whole series of distinctive
features. By the very fact that it is consensual, the contract is
covered by sanction, it is reciprocal and made in good faith.
That is not all. The principle on which the institution in its
new form rests, contains in itself the germ of a whole new
development. We must now trace the successive stages as well
as the causes, and determine its trends.

The consent may be given in ways that differ widely, depending on the circumstances, and so may exhibit qualities that
differ and that make it fluctuate in its value and moral significance. Granted that the consent was the basis of the contract, it
was natural that the public consciousness came to distinguish
the variations the consent could assume, to appraise them,
and so finally to calculate their legal and moral bearing.

The idea governing this development is that the consent is
truly itself, and binds truly and absolutely the one who consents,
only on condition that it has been freely given. Anything that
lessens the liberty of the contracting party, lessens the binding
force of the contract. This rule should not be confused with the

one that requires the contract to be made with deliberate intent. For I may very well have had the will to contract as I have done, and yet have contracted only under coercion. In this case, I will the obligations I subscribe to, but I will them by reason of pressure being put upon me. The consent in such instances is said to be invalidated and thus the contract is null and void.

To us this idea may seem a natural one but it only broke through very slowly and in the course of meeting with resistance of every kind. For centuries, the binding force of the contract had been supposed to reside outside the parties, in the formula pronounced, in the gesture made, in the thing delivered. Given this fact, the worth of the bond contracted could not be made to depend on what might have occurred in the depths of consciousness of the contracting parties, or on the conditions in which their resolve had been taken. It was not until the year 674 A.U.C., following on the dictatorship of Scylla, that an action was instituted in Rome to allow those who had been compelled under threats to contract undertakings injurious to them, to obtain compensation for this injury caused. It was the spectacle of the disorders and abuses witnessed in Rome under Scylla's reign of terror that suggested the idea. So this action came of a state of emergency but outlived it. It was given the name of *actio quod metus causa*. Its scope, by the way, was fairly limited. Let us take the case of a contracting party being in fear of a third party. That fear could only lead to an annulment of the contract if it was associated with some extreme form of evil and of a kind to shake the strongest. The only evils held to answer to this description were death or physical torture. Later, in milder times, other fears were added to that of death: these were the fear of arbitrary bondage, the fear of a capital charge or the fear of physical assault. Fears relating to honour or fortune were, however, never taken into account. (Accarias, *ibid.*)

In the present state of our laws, the rule has been still further mitigated. To invalidate a contract, the fear no longer has to be of a kind that only a stoic can stand up to it. According to the established formula (Art. 112), it is enough that the fear should affect any ordinary person. The text even adds that regard must be had "concerning this matter, to the age, sex and

station of the persons." The incidence of force or constraint
by fear is thus quite relative: in some cases it may even be very
slight. We have at last got away from the severe restraints of
Roman law.

What is the source of this legal precept, whose importance
we shall see directly? It is commonly said that man is a free
agent and that the consent he gives can be attributed to himself only on condition of its having been freely given. Here we
find ideas similar to those we meet with concerning responsibility. If the criminal has not committed an act of his own free
will, it is said, this act does not derive from him and he can
therefore not be blamed for it. It is the same with the contract.
There is a kind of responsibility, for instance, which arises from
a promise I have made, since I am bound to carry out certain
acts in consequence of this promise. But the other party to
whom it has been made can come to me with the demand it
be kept, only if it be really I myself who has made the promise.
If it has been imposed on me by a third party, it is not I myself,
in reality, who is responsible; therefore I could not be bound
by an undertaking that another has made, as it were, through
myself as intermediary. If the one who has used pressure on me
be also the one who benefits by the contract, he will have,
so to speak, no guarantor but himself; that is to say, such a
contract becomes null and void.

But this interpretation, to begin with, errs in setting the
legal system to solve problems in metaphysics. Is man a free
agent or is he not? That is a question which in fact has never
had any effect on legislation and it is easy to see why the law
ought not to hang on it. It is true we might think that the
state of public opinion on this controversial point could now
and then have contributed to deciding the spirit and letter of
the law in one way or another; we might hold the view that
public opinion changes according as the people believe in
freedom or not. The truth is that this question has never been
placed before the public consciousness in its abstract form.
Almost every society has believed in something that resembles
what is called freedom and at the same time in something that
corresponds to what is called determinism, without either of the
two concepts ever entirely excluding the other. From the advent
of Christianity onwards, for instance, we find at one and the

same time the theory of pre-determination by Providence, and the theory that holds every man to be the mainspring of his own faith and morality.

If man is a free agent, moreover, it seems as if he must always be in a position to refuse his consent if he so wills; hence, why should he not accept the consequences of that consent? The fact is all the more surprising and inexplicable since, in the case under review, of contract, quite slight acts of constraint are sometimes held to impair the consent. No great exertion is needed to resist that constraint. We do not allow a man to kill another to avoid some monetary loss and we make him responsible for his act. Still, nowadays we hold that the fear of an undeserved monetary loss is enough to invalidate a contract and to cancel the contracted obligations due from the one who has suffered this constraint. The freedom and power to resist are, however, the same in either case. How does it come about that in the one described the act should be regarded as done by free will and consent, and that in the other it assumes a quite different nature? Well, there are many instances in which the fear is intense and leaves room for no choice, and the will is therefore pre-determined, but in which nevertheless the contract is valid. The merchant who can escape the bankruptcy he is threatened with only by contracting a loan, has recourse to this means of saving himself because he cannot do otherwise; and yet, if the lender has not taken unfair advantage of the situation, the contract is valid morally and legally.

It is therefore not the amount of a greater or lesser freedom that matters; if contracts imposed by constraint, direct or indirect, are not binding, this does not arise from the state of the will when it gave consent. It arises from the consequences that an obligation thus formed inevitably brings upon the contracting party. It may be, in fact, that he took the step that has bound him only under external pressure, that his consent has been extracted from him. If this is so, it means that the consent was against his own interests and the justifiable needs he might have under the general principles of equity. The use of coercion could have had no other aim or consequence but that of forcing him to yield up some thing which he did not wish to, to do something he did not wish to do, or indeed of forcing him to the one action or the other on conditions he did not will.

Penalty and distress have thus been undeservedly laid on him. The feelings of sympathy that we usually have for our fellow-creatures are outraged when suffering is inflicted on someone when it is in no way deserved. The only kind of infliction that we find just is a penalty, and the penalty pre-supposes a culpable act. Any act must therefore seem immoral to us that causes injury to a fellow-man who has otherwise done nothing to alienate our ordinary human sympathies. We declare it to be unjust. Now an unjust act could not be sanctioned by law without inconsistency. This is why any contract in which pressure has a part, becomes invalid. It is not at all because the determining cause of the obligation is exterior to the individual who binds himself. It is because he has suffered some unjustified injury, because, in a word, such a contract is unjust. Thus, the coming on the scene of the contract by mutual consent, together with an increase in human sympathies, inclined the minds of men to the idea that the contract was only moral and only to be recognized and given sanction by society, provided it was not merely a means of exploiting one of the contracting parties, in a word, provided it was just.

What we should especially remember is that this principle was something quite new. This, in reality, is a new stage of the institution. The consensual contract pure and simple merely means, indeed, that the consent is the necessary but sufficing condition of the obligation. This new condition is now superimposed on the other which tends to become the essential condition. It is not enough that the contract shall be by consent. It has to be just, and the way in which the consent is given is now no more than the outward criterion of the degree of equity in the contract. The state or condition of the parties, taken subjectively, is no longer the single consideration. Now, it is only the objective consequences of the undertakings contracted that have a bearing on their worth. Put in another way, just as the consensual emerged from the ritual contract, so a new form succeeded to the consensual. This is the contract of equity, that is, objectively equitable. In the next lecture we shall see how this new principle developed and how, in developing, it was destined to have a profound effect on the present institution of property.

XVIII

MORALS OF CONTRACTUAL RELATIONS
(End)

IN the same way as the contract by mutual consent sprang from the ritual and the *real* contracts, so in turn did a new form begin to grow out of the consensual. This was the just contract, objective and equitable. Its existence was revealed when the rule appeared whereby the contract is null and void when one of the parties has given his consent only under pressure of obvious constraint. The society declines to approve a declaration of will which has been got only under duress. How does this come about? We have seen how slight the foundation is for attributing the legal consequences of constraint to the fact that it suppresses the free will of the agent. Should this word be taken in its metaphysical sense? Then, if man is a free agent, he is free to resist every kind of pressure exerted on him; his freedom remains unimpaired whatever the duress he may be exposed to. Is the meaning of an act of free will simply a spontaneous act and are we to understand that consent implies that the will in consenting does so spontaneously? How often it happens that we consent because we are tied by circumstances—compelled by them, without any option of choice. And yet, when it is things and not persons that exert this constraint over us, a contract made in these circumstances is binding. Under the pressure of illness, I have to call in a certain doctor whose fees are very high: I am just as much bound to accept them as if I had a pistol at my head. We might quote many other instances. There is always constraint in any acts we carry out and in any consent we give, for they are never exactly in line with our wishes. When we say contract we mean concessions or sacrifices made to avoid more serious ones. In this respect there are only differences of degree in the form that contracts take.

The true reason why contracts got under pressure have come to be condemned is that they cause injury to the contracting party who suffered the constraint. For it has compelled him to yield what he did not wish to, and it takes from him by force something that he owned. It is a case of extortion. What the law refuses to approve is any act having the effect of making a man suffer who might not have deserved it, that is, an unjust act. The law disallows this because the sympathy we all feel in our fellow-man rouses our opposition to suffering being inflicted on him; that is, unless he has earlier committed some act that dilutes our sympathy and may even turn it into antipathy. It is because the consent has been grievous in its effect that the society considers it null and void: that the individual has not, in a true sense, been the cause of the consent he has given is not the reason. And thus the validity of the contract becomes subordinate to the consequences it may have for the contracting party.

The injustices inflicted by constraint are, however, not the only ones that may be done in the course of contractual relations. They are only a variety. The one party, by knavery or excessive shrewdness or by knowing how to make adroit use of some unlucky turn in the affairs of the other, may bring him to consent to an exchange that is utterly unjust; that is, to consent to give his services or things he may own against a payment lower than their value. We know of course that in every society and in all ages, there exists a vague but lively sense of the value of the various services used in society, and of the values, too, of the things that are the subject of exchange. Although neither of these factors is regulated by tariff, there is, however, in every social group a state of opinion that fixes its normal value at least roughly. There is an average figure that is considered as the true price, as the one that expresses the true value of a thing at a given moment.

How this scale of values is arrived at is not for the moment our concern. All sorts of causes enter into the way it has evolved: that is, a sense of the true usefulness of things and services, of the labour they have cost, of the relative ease or difficulty in procuring them, traditions and prejudices of every kind, and so on. It remains true—and this alone matters to us for the time being—that this scale is certainly a real one, and that it is the touchstone by which the equity of the exchanges is to be judged. This normal price, of course, is an ideal price only: it very rarely coincides with the real price, which naturally varies according to circumstance; there is no official price-list to apply to every individual case. It is only a fixed point, around which there must inevitably be many fluctuations; but these cannot go beyond a certain range in any direction without seeming abnormal. We might even say that the more that societies evolve the more too does this structure of values become stable and regulated and unaffected by any local conditions or special circumstances, so that they come to assume an objective form. When every town and almost every village had its own market, the price scale varied according to the locality: each had the scale and tariff that suited it. These variations left far more lee-way to a shrewd personal ingenuity and calculation. This is why bargaining and individual prices are one of the characteristic features of petty trading and small-scale industry. The more we advance, on the other hand, the more do prices come to have an international basis: and this through the system of stock exchanges and controlled markets whose action covers a whole continent. Formerly, under the system of local markets, there had to be negotiating and a battle of wits, to know on what terms an object could be had; to-day, we only have to open a well-informed journal. We are becoming increasingly used to the idea that the true price of things exchanged should be fixed previous to the contract and be in no way governed by it.

Any contract, however, that diverges from these prices too greatly must needs seem unjust. An individual cannot exchange a thing for a price lower than its value without suffering a loss that cannot be made good or justified. It is just as if the amount unlawfully withheld were extorted under threat. In fact, we hold that there is a price that is due to him, and if he is denied it without cause our conscience rebels for the reason we mentioned before. The loss of standing inflicted on him, if he has not deserved it, wounds our sense of sympathy.

It hardly matters that he does not resist the indirect constraint put upon him and that he may even voluntarily accept it. There is something about this exploitation of one man by another that offends us and rouses our indignation, even if it is agreed to by the one who suffers it and has not been imposed by actual constraint. It is the same thing, of course, if the exchange is agreed at a price higher than the true value, for then it is the buyer who has been exploited. We see here that the notion of constraint recedes more and more into the background. (A just contract is not simply any contract that is freely consented to, that is, without explicit coercion; it is a contract by which things and services are exchanged at the true and normal value, in short, at the just value.)

Such contracts must seem to us immoral: no one will deny it. For contracts to be accepted as morally binding, we have come to require not only that they should be by consent, but that they respect the rights of the contracting parties. The very first of these rights is that things and services should not be given except at the fair price. We disapprove any contract with a 'lion's share' in it, that is, one that favours one party unduly at the expense of the other; therefore we hold that the society is not bound to enforce it or, at least, ought not to enforce it as fully as one that is equitable, since it does not call for an equal respect. It is true these views, with their source in the conscience, have so far remained moral ones and have not yet greatly affected the law. The only contracts of this sort that we absolutely decline to recognize are contracts of usury. Even here, the just rate, that is, the rate for lending the money, is fixed by law and may not exceed it. For various reasons we need not examine, this particular form of unjust exploitation is quicker to touch us and to rouse a deeper revulsion of conscience, perhaps because here the process is rather more physical and tangible.

Quite apart from the contract of usury, all regulations that are introduced in industrial law bear witness to the same need. These are designed to prevent the employer from abusing his position to get labour out of the workman on terms too much against his interests, that is to say, on terms that do not equate his true value. This is why we get proposals, whether justified of not, to fix a firm minimum wage. These are evidence that not every contract by consent is in our view one that is valid and just, even when there has been no actual coercion. In default of any regulations for a minimum wage, there are now provisions in the laws of several European countries that require the employer to insure the workman against sickness, old age and accidents. It was whilst this mood prevailed that our recent law was passed on industrial accidents. It is one of the many means employed by the legislative assembly to make the contract of labour less unjust. Wages are not fixed, but the employer is obliged to guarantee certain specific advantages to his employees. Protests are made and it is said this really amounts to giving privileges to the worker. In one sense this is quite true, but these are meant to counterbalance in part those other privileges enjoyed by the employer which leave him free to undervalue at will the services of the worker. I will not debate the usefulness attributed to these practices. It may be they are not the best or they may even work against the aim in view. No matter. It is enough to recognize the moral impulses that inspired them and whose reality they prove.

Everything goes to shew that we are not at the end of this development and that our demands on this score are rapidly growing. The feeling of human sympathy, indeed, which is their determining cause, is bound to gather greater force as it takes on a more egalitarian character. We are still inclined, under the influence of all kinds of prejudices inherited from the past, not to consider men of different classes from the same point of view. We are more sensitive to the distresses and undeserved hardships that a man of a superior class may undergo, who has important duties, than to the distress and burdens of those given up to humbler duties and labours. Everything leads us to suppose that this discrepancy in our way of sympathizing with different classes of people will tend gradually to fade away; that the misfortunes of one class will no longer seem more deplorable than the distresses of the other; that we shall consider them both as equally painful, since both are aspects of human suffering. Therefore we shall now be trying to take stronger measures to ensure that the contractual system shall hold an even balance between the two sides. We shall demand greater justice in contracts. I will not go so far as to say that the day will ever come when this justice will be absolute, when values will be exactly equated as between services exchanged. It might be said, and with reason, that it is not possible to carry it to the extreme limit. Are there not services which are beyond any adequate remuneration? Moreover, only a rough attempt can be made to make things square absolutely. But certainly, the balance of values that exists to-day still does not satisfy our present ideas of justice, and the more we advance the more we shall try to get near to the correct ratio. No one can set any limits to this development.

Now the supreme obstacle it comes up against is the institution of inheritance. It is obvious that inheritance, by creating inequalities amongst men from birth, that are unrelated to merit or services, invalidates the whole contractual system at its very roots. What indeed is the fundamental condition for ensuring the reciprocity of contracted services? It is this: for each to hold his own in this kind of duel from which the contract issues, and in the course of which the terms of exchange are fixed; the weapons of the contracting parties must match as nearly as possible. Then, and then alone; there will be neither victor nor vanquished; this means that things will be exchanged so as to balance exactly and to be equal in value. What the one receives will be equivalent to what he gives and vice versa. Conversely, a privileged contracting party could make use of the advantage he holds to impose his will on the other and oblige him to give the thing or service being exchanged at a price below its true value. If, for instance, the one contracts to obtain something to live on, and the other only to obtain something to live better on, it is clear that the force of resistance of the latter will far exceed that of the former, by the fact that he can drop the idea of contracting if he fails to get the terms he wants. The other cannot do this. He is therefore obliged to yield and to submit to what is laid down for him.

Now inheritance as an institution results in men being born either rich or poor; that is to say, there are two main classes in society, linked by all sorts of intermediate classes: the one which in order to live has to make its services acceptable to the other at whatever the cost; the other class which can do without these services, because it can call on certain resources, which may, however, not be equal to the services rendered by those who have them to offer. Therefore as long as such sharp class differences exist in society, fairly effective palliatives may lessen the injustice of contracts; but in principle, the system operates in conditions which do not allow of justice. It is not only to cover certain particular points that 'lion's share' contracts can be entered into, but the contract represents the 'lion's share' system as far as any relations of the two classes are concerned. It is the general lines on which the services of those not favoured by fortune are assessed that seem unjust, because the conditions stand in the way of their being reckoned at their true social value. The inherited fortune loads the scales and upsets the balance. It is in opposition to this inequitable assessment and to a whole state of society that allows it to happen, that we get the growing revolt of men's conscience. It is true that over the centuries, the injustice could be accepted without revolt because the demand for equality was less. To-day, however, it conflicts only too obviously with the attitude which is found underlying our morality.

We begin to appreciate what a signal event it was when this that we call the just contract came on the scene, and what widespread effects this concept was to have. The whole institution of property became transformed, since one of the sources of acquisition, and a principal one at that—I mean inheritance—stood condemned by this very concept. But it is not in this indirect and negative way alone that the development of contractual right tends to affect the right of property: that right is affected in a direct way. As we have said, justice demanded that services given or exchanged should not be remunerated below their value. This principle calls forth another, its corollary: that any value received must equate a service rendered. It is of course patent that in so far as the one value falls short of the other, the privileged individual can only have secured the excess value he may enjoy at the expense of someone else. This excess from which he benefits must have been the work of someone other than himself, who has been unlawfully deprived of it. If he is to receive more, that is, more than he is entitled to, another must receive less. Hence, we get this principle: the distribution of things amongst individuals can be just only if it be made relative to the social deserts of each one. The property of individuals should be the counter-part of the services they have rendered in the society. In this principle there is nothing that offends those humane feelings which are at the heart of this particular branch of morals. For this sympathy is liable to vary in depth according to the deserts of the individual as a social being. We have greater sympathy for those who serve the collectivity better and our goodwill towards them is all the greater; here nothing prompts us to protest if they are better treated—(with certain reservations we shall touch on). Again, a distribution of property on this pattern is closely in line with the interests of the society. For the society is concerned with seeing that things should be in the most capable hands.

The working of the principle, then, that lies at the foundation of the contract conforming to equity extends beyond the contractual right, and tends to become the basis of the right of property. As things are, the primary distribution of property is according to birth (institution of inheritance). The next stage is, that property originally distributed in this way is exchanged by contracts. But it is by contracts which, inevitably, are in part unjust as a result of an inherent state of inequality in the contracting parties, because of the institution of inheritance. This fundamental injustice in the right of property can only be eliminated as and when the sole economic inequalities dividing men are those resulting from the inequality of their services. That is why the development of the contractual right entails a whole re-casting of the morals of property. But close heed should be given to the way in which we summarily express this principle common to *real* right and contractual right. We are not going to say that property derives from labour, as if there were a kind of logical necessity for the thing to be attributed to the one who laboured to make it, as if labour and property were one and the same. There is nothing about the bond linking the thing with the person, as we described it, that can be analysed; there is nothing about labour that compels us to infer that the thing to which this labour has been applied derives from the workman. We have already shown all the unreason of such a deduction. It is the society that makes the synthesis of these two heterogeneous terms, property and labour. It is the society that does the allocation of property and it proceeds to allocate and distribute, according to the sentiments it has for the individual and moved too, by the way it calculates the value of his services. Since this way of calculating may be governed by principles that vary greatly, it follows that the right of property is in no wise something defined once and for all, a kind of immutable concept. No, on the contrary, it is something that can go on evolving indefinitely. Even the principle just mentioned can vary, more or less, and is therefore capable of developing. (We shall be reverting to this point.) At the same time, this is how we escaped the fallacies the classical economists and the socialists fell into, when they identified labour with property. Such an identification has a tendency indeed to make the output of labour take precedence over quality. But as we have said, it is not the amount of labour put into a thing which makes its value; it is the way in which the value of this thing is assessed by the society, and this valuation depends, not so much on the amount of energy expended, as on the useful results it produces, such at least as they are felt to be by the collectivity, for there is a subjective factor there which cannot be ruled out. An idea of genius, flowering without effort and created with joy, has greater value and merit than years of manual labour.

This agreed principle, although it is now graven in the conscience of civilized nations, is still not formally recognized by the law, and it raises a practical question. What reform could make it a reality in law? One primary reform is possible at once and almost without any transition. This is the discontinuance of inheritance *ab intestat* or by next of kin and above all of obligatory succession, allowed by our Code of Civil Law in the case of direct descent. We have seen moreover that inheritance *ab intestat*, a survival of the old right of family joint ownership, is to-day an archaic survival and without justification. It no longer corresponds to anything in our ethics and could be abolished without disturbing the moral structure of our societies in any way. As far as testamentary inheritance goes, it seems a more delicate matter. It is not because it is more easily reconciled with the principle we have raised. It offends the spirit of justice as much as inheritance *ab intestat* does and creates the same inequalities. Nowadays, we no longer allow a man to bequeath by will the titles or rank he acquired or the offices held in his lifetime. Why should property be any the more transferable? The position in society we have succeeded in attaining is at least as much our own creation as our fortune. If the law prohibits our disposing of the first, why should it be any different concerning the second—that is, property? Such a limitation to the right of disposal is in no way an attack on the individual concept of property—on the contrary. For individual property is property that begins and ends with the individual. It is the hereditary transference, whether by a man's Will or otherwise, that is contrary to the spirit of individualism. There are no real difficulties on this point, except when it is a question of testamentary inheritance in direct descent. Here a kind of conflict arises between our sense of justice and certain family customs that are very deeply rooted. It is clear that at the present day the idea that we could be prevented from leaving our possessions to our children would meet with very lively resistance. For our work is done quite as much to ensure their happiness as our own. That does not mean that this state of mind does not derive very closely from the present structure of property. Let us grant that there is a transfer by inheritance and in consequence an initial inequality in the economic status of individuals at the time they enter the life of the society. We then attempt to make this inequality have as little disadvantage as possible for the human beings with whom we have the closest ties; we go further, and try to make it even a positive advantage. Hence our anxiety to work for them. But if equality were the rule, this need would be of far less concern to us. For the peril to them of facing life with no resources but their own would have disappeared. This peril comes solely from certain people being at present endowed with initial advantages, a fact that places those not so endowed in a position obviously inferior. All the same, it is not unlikely that something would always remain of the right to dispose of property by will. The old institutions never disappear entirely; they only pass into the background and fade away by degrees. This one has played too great a role in history for it to be conceivable that nothing of it should survive. It would only survive, however, in a weakened form. We might for instance imagine that every head of a family would have the right to leave to his children specified portions of the heritage. The inequalities that would then continue would be so slight as not to seriously affect the working of the contractual right.

And so, it is beyond us to make any very accurate forecast on this subject, for one factor needed in making it is at present lacking. To whom, indeed, would the wealth go to that each generation would leave behind without an owner as it left the scene? When there were no longer any heirs either by birth or by right, who would then inherit? The State? It is clearly impossible to concentrate such vast resources in hands that are already so blundering and wasteful. Alternatively, a periodic sharing-out of these things amongst individuals would have to be made, or at the very least of certain things, such as those essential to labour, of the land, for instance. Surely we can imagine some form of auction, when things of this kind would be knocked down to the highest bidder. But it is obvious that the State is too far removed from things and individuals to be able to carry out tasks so vast and so complex with any competence. There would have to be secondary groups, more limited in range and closer to the facts in detail, to be able to fulfil this function. We could hardly choose any better suited to the task than the professional groups. They are well equipped to manage any particular set of interests and could branch out into all parts of the country; at the same time they would take into account the regional differences and purely local affairs. They would satisfy all the conditions for becoming in a sense, in the economic sphere, the heirs of the family.

The family was in the past better suited to ensure the continuity of economic life as well, because it was a small group in direct touch with things and people and also itself endowed with a genuine continuity. To-day this continuity no longer exists. The family is all the time in process of breaking up; it lasts only for a period and it may die out here and there. It no longer has sufficient power to link the generations one to another, in the economic sense. But only a secondary, fairly small medium can be a substitute. This can and should have greater scope than the family because the economic interests themselves have grown in importance and are found to touch every part of the country. It is not possible for any central organ to be everywhere present and everywhere active at the same moment. All these points, then, persuade us in favour of the professional groups.

Beyond these practical conclusions, this study of contractual right leads us to put forward an important point of theory. In the sphere of ethics we have just been examining, that is, in the morals of human behaviour, we usually distinguish between two very different varieties of duty. In one, they are known as duties governed by justice and in the other, they are duties governed by charity. Between the two, it is agreed there is a kind of hiatus or break in continuity. They seem to derive from ideas and sentiments that have nothing in common. In justice, there is a further division as between distributive justice and commutative justice. The second of these is the justice that governs or ought to govern exchanges, to the end that we receive always a just recompense for what we give. The first relates to the way the laws are applied and office and rank allocated or shared out by the society amongst its members. The result of all this is that there are only differences of degree amongst these various layers of morals and that they are in line with one and the same collective consciousness and with one and the same collective sentiment taken at different periods in the development of these.

To begin with, as far as a distributive justice and a commutative justice are concerned, we have seen that they mutually affect one another and are mutually involved. If exchanges are to be equitable, they have to be justly balanced, and of course the distribution of things, even if it had followed all the rules of equity to begin with, would still not remain just, if exchanges could be contracted on unjust terms. Both are the sequel in law of the same moral sentiment—the sympathy that man has for man. It is only that this sympathy is considered in both cases from two different aspects. In the one, the feeling is against the individual giving more than he receives or rendering services that are not rewarded at their true value. In the other case, this same feeling requires that there shall be no social inequalities as between one man and another, except those that reflect their own unequal value to society. In a word, this sentiment, in both its aspects, tends to eliminate or strip away from all social sanctions every kind of physical and material inequality—all inequalities that derive from the accident of birth or from family status, leaving only those of merit.

If justice alone is in question, these inequalities of merit will still persist. But where human sympathy is concerned, even these inequalities can not be justified. For it is man as a human being that we love or should love and regard, not man as a scholar of genius or as an able man of business, and so on. . . . Essentially, are not these inequalities of merit fortuitous, too?

219

For these all men are born with—by temperament, and it seems hardly just to make them bear responsibility for them. To us it does not seem equitable that a man should be better treated as a social being because he was born of parentage that is rich or of high rank. But is it any more equitable that he should be better treated because he was born of a father of higher intelligence or in a more favourable moral *milieu?* It is here that the domain of charity begins. Charity is the feeling of human sympathy that we see becoming clear even of these last remaining traces of inequality. It ignores and denies any special merit in gifts or mental capacity acquired by heredity. This, then, is the very acme of justice. It is society, we find that is coming to exercise complete dominion over nature, to lay down the law for it and to set this moral equality over physical inequality which in fact is inherent in things.

However, we see that this feeling of human sympathy only comes to have this depth in some rare forms of consciousness, the highest; consciousnesses remain as a rule too feeble to go the whole way in their logical development. We have not yet reached the day when man can love all his fellow-creatures as brothers, whatever their faculties, their intellect or their moral values. Nor has man reached the stage when he has shed his egotism so successfully that it is no longer necessary to put a tentative price on merit (a price likely to decline), with the purpose of stimulating it (the merit) and of keeping the price steady. This is what makes a complete levelling to equal values impossible to-day. On the other hand, it is certain that the depth of feeling of human fraternity will go on increasing, and that the best amongst men are capable of working without getting an exact recompense for their pains and their services. This is how it comes about that we go on trying to soften and tone down the effects of a distributive and commutative justice which are too strictly reckoned, which, that is, in reality remain unachieved.

This is why, as we go on, charity, in its true meaning, becomes ever more ?significant (illegible) and so it ceases, as it were, to be optional and to go beyond what it need be, and becomes instead a strict obligation, that may be the spring of new institutions.

INDEX

Index

Contract
feudal, 188–9
formula, 182, 186–7, 190–91, 202
freedom of, 67–8
labour, 123, 211–12
law of, 181
  French, 204–05
  Germanic, 200
  Roman, 187, 195, 199–200, 201, 202
' lion's share ', 211, 213–14
loan, 201
marriage, 177, 180, 189, 195
morals of contractual relations, 184 ff
nature of, 176 ff
oath, 182, 186–8, 190–191, 193–4, 196
origins of, 177 ff
penal sanctions, 198
processes, xl–xli
promise of parties, 182
property transfer, 173
quasi-, 175
real, 180, 181, 183, 188–90, 192, 193, 196–7, 199, 201, 202
right of, 169–70, 171 ff, 179, 196 ff, Ch. xvii
rights and duties, 185
' ritual ', *see* Contract, solemn
Rituals, 187–91, 193–5
sacramentum, 182–3
sacred origins, 178, 182–3
social, 175 ff
solemn, 182, 186–8, 192–4, 196–7, 199–202
in trade transactions, 191
unilateral, 184
usury, 211
will of parties to, 178–9, 181–6, 192, 193, 194, 201–03
words in pledges, 186, 202–03
Contrat Social, 135
Copper-smiths, guild, 32
Corporate bodies,
  see Corporations,
  see also Guilds,
' *Corporation* ', footnote 17
Corporations, compulsory membership, 39
definition, 8
as the electorate, 105
employer and employed, 39
functions, 40
as intermediary electoral units, 103–04, 106

Corporations
social value of, xxxv
and the State, 40
as successors to Guilds, 31, 37–41
see also Guilds,
         Groups, Professional,
Corsica, homicide, 116
Council of Ministers, 51
Craft guilds, } *see* Guilds
  unions, }
Craftsman, the
and the State, 64
Crime,
and collective consciousness, xvii
nature of, 110 ff
sanctions, 111
statistics, 117
Criminology, 110
Crozat, Charles, xi
Customs, in making of contracts, 189–90
origins and symbolism, 189
relating to debt, 198–9
Cutlers, guild, 23

Dead, the,
cult of, 153
customs and beliefs, 198–9
Debt, penalties, 198–200
Democracy, 76 ff, 85 ff, 98 ff
  pseudo-, 84, 88
Denier à Dieu, (contract), 187
Descartes, xix
Despotism, 76
Determinism, 205–06
' Dharna ',
debt penalty, 198, 199
Dionysius of Halicarnassus, 32
Diplomacy,
and the State, 85
Division du Travail Social, xvi, xix–xx, xxxvi
Dolo, in Roman law, 202
Domestic group
  see Family
Domestic society, definition, xxxvii
Donations *inter vivos*, 123, 124, 140
Duties, governed by charity, 218–19
justice, 218–19
moral rule, 3–5
of various groups, 3–6

222

Index

Index

Guilds
regulations, 22–3, 28–9
Roman, 19–22, 31–2
trade, 18
traditionalism, 38
see also Corporation
Groups, professional

HALPHEN, Madame Jacques, x
Hegel, 54
Hindu Law, 198
History, interpretation by, xxxiii
philosophy of, xxxi, xxxiii
Homicide, 110 ff
statistics, xix, 113–14, 119
Honolulu, taboo, 144
Hornblowers, guild, 32
House building, ritual, 152, 157
Humane sciences, xiv
Hungary, 114, 119

IDEALS, source of, xliii
Ideas, science of, xiv
Ihering, 58
' Imperative mandate '
of the electorate, 91
Incest
and property, 177
India, 150, 152, 198
Individual, the
autonomy of, 91
collective life, 90
and the family, 4
freedom of, xxxix
and fruits of his labour, 122–23
and the law, 107
as member of group, 14–15
morals and, xlii, 3 ff, 14–15, 42 ff,
112
and religion, 55–6
religious cult of, 69–70
respect for, xxxix, 172
rights of, 52–3, 57, 60, 65, 68
social dimension, xli
in social groups, xxxvi
and the State, xxxviii, xxxix, 42,
54, 55 ff, 65 ff
see also Person
Individual consciousness
see Consciousness,
individual and collective
Individualism,
in the City States, 59

in France, 60
and religion, 58–9
Industrial law
employer and employed, 211
Industry
grouping by categories, 37–9
health, 40
large-scale, 35, 36, 37, 38–9, 41
legislation, 40
and professional ethics, 9–10
Inheritance, 123,
ab intestat, 216
and contractual system, 213–18
rights, 40
State as heir, 218
testamentary, 216, 217
see also under Property
Ireland, 198
Italy, 44, 114, 119

JACQUARD machines, 126
Japan, 198
Judaea, crimes in, 111
Jupiter, Capitoline, 151
Juridical facts, 2
Jus abutendi, 138–41
Jus civilis, 34
Jus fori, 34
Jus fruendi, 138–41
Jus utendi, 138–41
Justice, and charity, 220
commutative and distributive, 219,
220
prohibitive, 64, 65
Justinian, 137, 150

KANT, 51, 52, 64, 65, 66, 68, 69, 119,
127–36
Kinship, 25
and property, 164
Koral, M. Rabi, xi
Kubali, H. N., ix

LABOUR
assessment of value, 216
contract, 40, 66, 67, 123, 169,
173–4, 211–12
disputes, 40
division of, xxxvi, 17–18
*see also Division du
Travail Social*
and property, 121–5, 131, 135,
215–16
remuneration, 66, 67

Index

Index

Index